Microbial Ecology
Microbiomes, Viromes, and Biofilms

Editor

Bhagwan Narayan Rekadwad

Associate Professor
Microbe AI Lab
Division of Microbiology and Biotechnology
Yenepoya Research Centre
Yenepoya (Deemed to be University)
Mangalore, Karnataka
India

CRC Press is an imprint of the
Taylor & Francis Group, an **informa** business
A SCIENCE PUBLISHERS BOOK

Cover credit: Images reproduced by kind courtesy of the editor.

First edition published 2025
by CRC Press
2385 NW Executive Center Drive, Suite 320, Boca Raton FL 33431

and by CRC Press
4 Park Square, Milton Park, Abingdon, Oxon, OX14 4RN

CRC Press is an imprint of Taylor & Francis Group, LLC

Library of Congress Cataloging-in-Publication Data (applied for)

ISBN: 978-1-032-50605-0 (hbk)
ISBN: 978-1-032-50606-7 (pbk)
ISBN: 978-1-003-39924-7 (ebk)

DOI: 10.1201/9781003399247

Typeset in Palatino Linotype
by Prime Publishing Services

Preface

The study of microbial ecology is a rapidly advancing field shedding new light on the complex relationships between microorganisms and their environments. At the heart of this field are three key concepts: Microbiomes, viromes and biofilms. Microbiomes refer to the communities of microorganisms that live on or within a particular host or environment, while viromes represent the viruses that infect these microbial communities. Biofilms are complex communities of microorganisms that attach to surfaces and form intricate structures. Understanding the roles that these microbial communities play in shaping human health and disease is a major focus of this book. Microbial infections are a major cause of morbidity and mortality worldwide. Apart from basic/fundamental microbiological analysis; microbiomes, viromes and biofilms have been linked to various diseases, including diabetes mellitus and liver diseases. For example, the gut microbiota plays a crucial role in metabolic processes and immune function, and disruptions to this community have been linked to the development of diabetes mellitus. Similarly, changes in the gut microbiota have been implicated in the development of alcoholic and non-alcoholic fatty liver diseases. This book explores the latest research on the role of microbiomes and viromes in infections, providing insights into how we might prevent and treat these diseases. Microbiomes, viromes and biofilms are not only associated with infections but also with human traits and environmental conditions. For example, the gut microbiota has been linked to mood and behaviour, and changes in this community have been implicated in the development of depression and anxiety.

Additionally, microbiomes, viromes and biofilms play critical roles in environmental processes, including nutrient cycling and bioremediation. This book explores the many ways in which microbiomes and viromes influence human health and the environment. Probiotics and other interventions that manipulate the microbiome and virome have become increasingly popular in recent years. However, the effectiveness of these interventions remains unclear, and there is much debate about the appropriate use of such therapies. This book provides an in-depth examination of the microbiome and probiotics, exploring the evidence for their efficacy and discussing their potential applications in various disease states. Hence, the use of next-generation sequencing and bioinformatics tools has revolutionised our ability to study microbiomes, viromes and biofilms. These tools allow us to identify and characterise microorganisms with unprecedented accuracy and speed, providing new insights into the roles that these communities play in shaping our world. This book provides a comprehensive overview of these technologies, offering a practical guide on their use in microbiome, virome and biofilm research.

Dr. Bhagwan Narayan Rekadwad

Contents

CHAPTER 1

Introduction of Microbiomes, Viromes and Biofilms

Murad Muhammad,[1,3] *Wen-Jun Li,*[1,2,*] *Li Li,*[1]
Yong-Hong Liu,[1] *Kashif Ali*[4] and *Iftikhar Ahmed*[5]

Introduction

One of the hardest things about ecology is trying to figure out what controls the number and variety of species, how these things are affected by natural or human-caused changes, and how these changes manifest in the processes and features of an ecosystem. These are some of the most critical questions that ecology can address. Globalization and climate change facilitate species' mobility between locations, consequently leading to an increase in the prevalence of invasive species. However, significant uncertainties persist regarding the capacity of humans to detect and predict biological invasions (Van der Putten et al. 2007). The biological charges of

[1] State Key Laboratory of Desert and Oasis Ecology, Key Laboratory of Ecological Safety and Sustainable Development in Arid Lands, Xinjiang Institute of Ecology and Geography, Chinese Academy of Sciences, Urumqi, 830011, PR China.

[2] State Key Laboratory of Biocontrol, Guangdong Provincial Key Laboratory of Plant Resources and Southern Marine Science and Engineering Guangdong Laboratory (Zhuhai), School of Life Sciences, Sun Yat-Sen University, Guangzhou, 510275, PR China.

[3] University of Chinese Academy of Sciences, Beijing 100049, China.

[4] Center of Biotechnology and Microbiology, University of Peshawar, Peshawar 25120 Pakistan.

[5] National Culture Collection of Pakistan (NCCP), Bio-Resources Conservation Institute (BCI), National Agricultural Research Centre (NARC), Park Road, Islamabad 45500, Pakistan.

* Corresponding author: liwenjun3@mail.sysu.edu.cn

exotic species into terrestrial ecosystems are among the most critical threats to the native biodiversity and ecological stability of these ecosystems. (Sousa et al. 2011). When discussing ecosystems, people often focus on invasive plants and other animals that dwell above the ground (Bardgett and Wardle 2010). However, the visible biota that inhabits ecosystems can significantly impact and move the unseen microbiological components of ecosystems and the processes that are driven by those components. In addition, there is an increasing realization that invasive microbes, such as those that cause diseases in humans, animals or plants, can change the structure and function of entire ecosystems. These diseases can affect humans, animals, or plants (Litchman 2010). The microbial communities found in the environment are immensely diverse and complicated. Some examples of these communities include microorganisms such as bacteria, archaea, fungi, and viruses. The intricate networks of connections created by these bacteria significantly impact the functioning of ecosystems and the biotic and abiotic components that make up an ecosystem (Shade et al. 2012).

This chapter will focus on three essential aspects of microbial ecology: microbiomes, viromes, and biofilms. Microbiomes - colonies of bacteria – are found in environments, and these microbiomes are diverse and dynamic. The billions of bacteria that inhabit the human body have co-evolved with us, resulting in highly specialized adaptive ecosystems that are precisely tuned to the continuously fluctuating physiology of the host. Because of this, the human body is home to various microbes (Jenkins 2019). Cancer, Asperger's syndrome, inflammatory bowel disease, multiple sclerosis, allergies, asthma, and type 1 and type 2 diabetes are just a few of the disorders linked to microbiome dysbiosis. Determining the pathogenicity of a single microbial taxon or the dysbiosis of an entire microbial community might be challenging. Dysbiosis may be understood as a deviation from a healthy ecology that serves to prolong, exacerbate, or generate an unfavorable health outcome (Decker et al. 2021). Finding new approaches to prevent diseases or improve prognosis may depend on uncovering the characteristics that distinguish healthy microbiomes from bad microbiomes, which could aid in diagnosing microbiome-related disorders. Healthy microbiomes are thought to share certain features, including common species and metabolic activity (Lloyd-Price et al. 2016). According to Willis and Gabaldón's 2020 research, the human microbiome, much like the earth's multiple terrestrial biomes, is composed of a varied collection of microbial communities, which changes depending on the surrounding environment. It is possible to think of different body sections as different biomes, each with a unique atmosphere and availability of various nutrients – both encouraging their unique community of cells. The makeup of a microbiome might vary substantially (even at the level of a single body site) across persons whose health and lifestyles are drastically different from one another (Porras-Alfaro and Bayman 2011). The viruses included in a virome are numerous and diverse and come from various kingdoms of life. The human body is home to various eukaryotic and prokaryotic viruses, collectively comprising the human viromes. However, since every organ and tissue in the body has its particular microenvironment, the viromes might change depending on where they are situated. The makeup of these viromes can also be influenced by factors such as age, diet, and the presence or absence of specific microbiome components (Liang and Bushman 2021).

Biofilms are formations that are both dynamic and complex. They are produced by communities of microorganisms that live close to one another and adhere to surfaces. Communities of bacteria attached to surfaces can be referred to as biofilms. Cells that constitute a biofilm can be differentiated from their counterparts suspended in the medium by producing an extracellular polymeric substance (EPS) matrix, slowing their growth rates, and up-and-down-regulation of specific genes. The regulation of the attachment process is affected by various elements found on the cell surface, as well as those in the growth medium and the substratum (Kokare et al. 2009). A developed biofilm structure comprises microbial cells and extracellular polymeric substances (EPS), has a unique structure, and provides an excellent environment for cells to trade DNA with one another. Cell-to-cell communication in quorum sensing may regulate biofilm activities such as detachment (Dewasthale et al. 2018).

The advancements that have been made in next-generation sequencing over the last decade have led to a significant expansion of our understanding of the human microbiome and the role it plays in both health and disease (Malla et al. 2019). However, our knowledge of the human virome is still quite limited, particularly in how it interacts with essential microbes that affect human health. The human body is home to many prokaryotic and eukaryotic viruses, each contributing to maintaining a distinct niche and having various consequences on our health (Zárate et al. 2017). Viruses may be divided into two categories: those that are prokaryotic and those that are eukaryotic. Although phages and other prokaryotic viruses have been discovered in different regions of the body, the digestive tract of a human being provides an especially favorable environment for the growth of these viruses (Hidalgo-Cantabrana et al. 2018). Because of the matrix in which they are encased, bacteria that generate biofilms are protected from various stimuli, including antibiotics, ultraviolet light, chemical biocides, the immunological response of the host, and other irritants (Berhe et al. 2017). As biofilms provide a protective coating, microorganisms can endure adverse conditions such as high salinity and pressure, extremes in temperature and pH, insufficient nourishment, antibiotics, etc. (Yao et al. 2022). Most antibiotic resistance may be traced back to structural barriers and the persistent cells found in biofilms. According to research, infections brought on by biofilms are notoriously difficult to treat (Abdelghafar et al. 2022). Antibiotics are no longer effective in treating conditions caused by biofilms due to the development of antibiotic resistance and genetic mutations. Biofilms, which have recently been identified as a significant factor in causing illnesses not cured by antibiotics, are frequently the source of antibiotic-resistant bacteria (Bowler 2018). It is common knowledge that biofilms are responsible for more than 80 percent of all chronic infectious disorders. It is also common knowledge that conventional antibiotic therapies cannot address infections mediated by biofilms and can not cure them (Li and Lee 2017). Consequently, the healthcare business faces a considerable risk from the presence of bacteria that create biofilms. This chapter aims to provide a brief summary of what is currently known about microbiomes, viromes and biofilms, along with their significant contributions to and interactions with microorganisms and their environment, as well as the challenges presented by this rapidly growing field and the opportunities it presents.

Microbiome Diversity

Throughout millions of years, microorganisms on Earth have co-evolved with other species to become highly specialized contributors to ecological communities. Microbiomes are communities of microorganisms that can live in and on a host organism for an extended period (Foo et al. 2017). The microbiome is a complex ecosystem of microorganisms found in or on the human body and includes microbes such as bacteria, viruses, fungi, and protozoa (Davis 2016). Owing to its vital role in maintaining human health and well-being, research on microbiomes has grown increasingly prevalent in recent years. One of the essential elements of a microbiome is its diversity, which can be defined as the wide range of microorganisms found in a specific habitat (Kim et al. 2017).

About ten percent of the cells that make up our bodies originate from the human host; the remaining cells come from the human microbiota. These commensal microbes help us resist diseases, train our immune systems, and impart some characteristics that humans did not initially evolve within their bodies (Byrd et al. 2018). For instance, we frequently ingest plant polysaccharides that are abundant in carbohydrate structures containing xylan, pectin, and arabinose. Even though human DNA lacks most of the enzymes necessary for decomposing these chemicals, the microbiota in our distal gut allows us to do so (Agans et al. 2018). The human genetic landscape combines the human genome and the metagenomes of microbes that have colonized human bodies (Smith and Wissel 2019). Because of this, the genetic diversity of humans is not only found in the allele frequencies of shared *Homo sapiens* genes but also the genes included within our various microbial communities (Zilber-Rosenberg and Rosenberg 2021). To understand the genetic and physiological differences observed, it is necessary to characterize the composition and structure of human microbiota in significant regions of the body (such as the mouth, skin and gut), as well as the variables that influence them.

Oral Bacterial Microbiome

The oral cavity is home to about 700 unique species of bacteria, making it the second most varied microbiota after the digestive system. In addition to bacteria, fungi, viruses, and protozoa, it fosters the growth of many other microorganisms and is a favorable environment for their survival (Wiley et al. 2017). Tooth decay, also known as dental caries, and gum disease, sometimes periodontal disease, are the two most frequent bacterial diseases that affect humans (Mosaddad et al. 2019). The mouth is a breeding ground for a wide variety of bacterial species. It is consistently exposed to environmental factors, and research has revealed that it is susceptible to the effects of these influences. The human microbiome has a relatively stable core and a more fluid peripheral. The core microbiome of a body in good health comprises the most prevalent species at each location. Since it has grown in reaction to a person's distinctive genetic make-up and how they live their life, each individual's microbiome is unique (Balan et al. 2017). Approximately 700 unique types of prokaryotes have been discovered. Around 54% of these species have been given their proper names, 14% are not named but are cultivated, and 32% are only known as uncultivated phylotypes. These species are divided into 185 genera and 12 phyla (Zhao et al. 2017b). The following phyla and orders of oral bacteria are found most frequently in humans (Table 1).

There is a consistent microbial community in healthy mouths, down to the genus level in the oral microbiome. Despite the similarities, the variety of a person's microbiome is highly dependent on their geographical location. Microflora, some of which are anaerobes, can be found on the tongue, thanks to its many papillae, despite its relatively low number of anaerobic sites. A limited variety of microorganisms is present in the buccal and palatal mucosae (Sultan et al. 2018). The most prevalent bacteria

Table 1. Major phyla and class of human oral microbiota (Idris et al. 2017).

Phylum	Class
Proteobacteria	*Alphaproteobacteria* *Epsilonproteobacteria* *Deltaproteobacteria* *Gammaproteobacteria* *Betaproteobacteria*
Chloroflexi	*Caldilineae* *Anaerolineae*
Gracilibacteria (GN02)	*GN02 [C-2]* *GN02 [C-1]*
WPS-2	*WPS-2 [C-1]*
Saccharibacteria (TM7)	*TM7 [C-1]*
Actinobacteria	*Actinobacteria*
Firmicutes	*Erysipelotrichia* *Negativicutes* *Mollicutes* *Bacilli* *Clostridia*
SR1	*SR1 [C-1]* *SR1 [C-2]* *SR1 [C-3]*
Bacteroidetes	*Flavobacteriia* *Bacteroidia* *Sphingobacteriia* *Bacteroidetes [C-1]* *Bacteroidetes [C-2]*
Spirochaetes	*Spirochaetia*
Chlamydiae	*Chlamydiia*
Synergistetes	*Synergistia*
Fusobacteria	*Fusobacteriia*
Chlorobi	*Ignavibacteria* *Chlorobia*

that can be discovered in a healthy mouth are listed below (Amaroli et al. 2022).

Gram-positive:

1. **Rods:** *Bifidobacterium, Propionibacterium, Eubacterium, Lactobacillus, Pseudoramibacter, Actinomyces, Rothia, Corynebacterium.*
2. **Cocci:** *Abiotrophia, Peptostreptococcus, Streptococcus, Stomatococcus.*

Gram-negative:

1. **Rods:** *Campylobacter, Leptotrichia, Hemophilus, Desulfobacter, Capnocytophaga, Simonsiella, Desulfovibrio, Eikenella, Fusobacterium, Treponema, Prevotella, Selemonas, Wolinella.*
2. **Cocci:** *Neisseria, Moraxella, Veillonella.*

In addition to bacteria, the mouth is home to a wide variety of other organisms, including protozoa, fungi, and viruses; *Entamoeba gingivalis* and *Trichomonas tenax* are the protozoa that are most commonly encountered. Both species are predominantly saprophytic and pose no threat to human health (Deo and Deshmukh 2019). Fungi of many different kinds can be discovered in and around the oral cavity, but the *Candida* species is by far the most common. In culture-independent studies conducted on twenty more healthy hosts, researchers detected a total of 85 unique species of fungi. It has been concluded that the bulk of the detected species might be attributed to the genera *Candida, Cladosporium, Aureobasidium, Saccharomycetales, Aspergillus, Fusarium,* and *Cryptococcus* (Sharma et al. 2018).

Skin Microbiota

The skin is the body's largest organ and its primary protection mechanism against external environments (McLoughlin et al. 2022). There are millions of different kinds of bacteria, fungi, and viruses in the skin's microbiome. Like the microorganisms in the digestive tract, those on the skin also perform essential roles, including defense against pathogens, immune system training, and waste disposal (Falcao and Inaturals 2021). Reduced diversity of gut microbiota compared to cutaneous microbiota is possible. According to Lima et al. (2020), the vulnerability of this microbial network lies in the multiple intrinsic and extrinsic components that affect it (Figure 1). The connections between these two areas highlight the value of skin homeostasis in treating wounds and preventing complications from infections or the effects of harsh environments. Recent research has found a correlation between alterations in these mutually beneficial populations and physiological changes, such as aging and other skin disorders, in animals and humans (Nguyen and Kalan 2022).

Gut Microbiota

The term 'gut microbiota' refers to the diverse population of bacteria, archaea, and eukaryotic organisms that inhabit the digestive tract. The

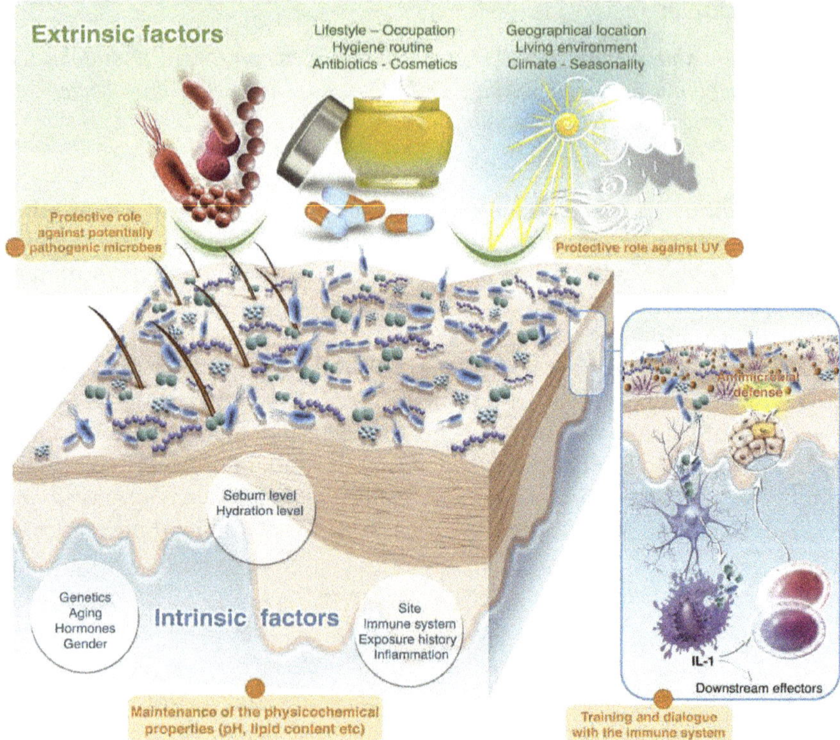

Figure 1. The elements that shape the microbiota that lives on or within the human skin. The microbiota that dwells on or in the skin is formed by both external (such as lifestyle, which includes work, hygiene practices, and drug and cosmetic usage) and internal (such as inheritance) influences (genetics, aging, sex, site of the body, etc.). Protecting the skin from potential infections or climatic changes and maintaining the skin's integrity are only two of the many ways these factors might affect the functioning of the skin's microbiota (Boxberger et al. 2021).

human host and its accompanying bacteria have co-evolved in a complex and mutually beneficial relationship over millions of years (UMAR et al. 2022). It has been estimated that more than 1,014 types of bacteria dwell in the human digestive system. There are around ten times as many bacterial cells as human cells, and the microbiome has over one hundred times the genomic information in the human genome. A recent study, however, indicates that the ratio of human to bacterial cells is closer to one-to-one. The combination of a host and its microorganisms is frequently called a 'superorganism.' This happens because the human body contains many bacterial cells (Thursby and Juge 2017).

The gut microbiota has been linked to numerous human diseases, including luminal disorders such as inflammatory bowel disease (IBD) and irritable bowel syndrome (IBS), metabolic diseases such as obesity and diabetes, allergic diseases, and neurodevelopmental disorders. However, the evidence supporting many of these claims is scant. In recent years, scientists have focused much attention on gut microbiota (Gomaa 2020). It has been hypothesized that the microbiota in the gut plays a crucial role, both structurally and functionally, in the maintenance of the gut in healthy individuals and human health as a whole (Sinha et al. 2023, Makki et al. 2018).

The host's immune system can recognize infections and attempt to eliminate them. Immunology serves as the basis of the immune system, which explains why this is the case. Yet, most gut bacteria are harmless and connect positively with enterocytes (Kundu Smita and Rana 2021). In the digestive system, commensal bacteria are essential in metabolizing nutrients and medicines, preventing infections, and maintaining intestinal barrier function. In the interim, the immune system has co-evolved to collaborate with microbiota that is helpful to the host while protecting the host against pathogenic microorganisms. As a result, the immune system can perform both roles effectively (Kogut et al. 2020).

A person's genetics, nutrition, lifestyle, age, and environment influence the variety of the microbiota in their gut (Rinninella et al. 2019). The foods we eat affect the bacteria that inhabit our gastrointestinal system. Compared to a diet rich in fiber, fruit, and vegetables, a diet rich in fat and sugar results in an imbalanced gut flora. In contrast, a diet rich in fiber, fruit, and vegetables promotes the growth of beneficial microorganisms (Luo et al. 2021). This microbial population may experience a decline in variety due to the use of antibiotics and other drugs, which can alter the composition of the gut microbiota (Barko et al. 2018).

Virome

Infectious agents known as viruses can infect living things throughout all kingdoms of life. They frequently infect new hosts and occasionally produce diseases that incapacitate the host (Chevallereau et al. 2022). The virome contains the world's most abundant and rapidly changing genetic components. It is also known as the viral genome. The term 'mammalian virome' refers to the collection of viruses that infect host cells, chromosomal segments produced from viruses, and viruses that infect the wide variety of other organisms that live in and on mammals. Chromosomal components that are made from viruses are also included in this collection. Evidence shows these viruses have contributed to the mammalian virome (Santoro et al. 2020). The microbiome comprises all the bacteria that reside inside and on our skin, and the virome is a subset of the microbiome. The collection of all of these bacteria is known as the microbiome. This group includes bacteria and archaea, fungi, protozoa, and any other meiofauna that can be present (Cadwell 2015).

The virome has only recently become 'visible' in large sequence datasets, thanks to developments in next-generation sequencing and bioinformatics that make it possible to detect links between viruses despite significant nucleotide sequence variation. This has been made possible by the fact that we can now see viral links even though nucleotide sequences can vary greatly. As a direct result, research on the virome is still in its infant stages. The virome has recently become 'visible' in big sequencing datasets (Esposito et al. 2022). The mammalian virome contains a variety of viruses, including eukaryotic viruses (eukaryotic virome), bacteriophages (bacterial virome), archaeal viruses (archaeal virome), and virus-derived genetic elements (VGEs) on host chromosomes. All of these viruses (or virus components) can alter the gene expression of their hosts, express proteins, and even produce infectious virus particles (pro-phages, endogenous retroviruses, endogenous viral elements). Currently, we have the least information on the archaeal virome, which refers to the viral population that can be detected in human cells (Zhao et al. 2019). Despite the considerable variations between eukaryotic and bacterial viruses, both utilize 'lytic' and 'latent' life cycles. The host cell is killed during viral replication, and the virus remains dormant within a living cell. The host cell is destroyed during viral replication in lytic life cycles; in latent life cycles, the virus remains dormant within a living cell. The viral genome can remain inactive in the host cell as an episome (as with herpesviruses) or be incorporated into the host's chromosome (as with prophages) until it becomes more infectious. This phenomenon is seen in the case of herpesviruses. Herpesviruses are an excellent example of this behavior in the natural world (Legoff et al. 2020). Because of the diverse range of lifestyles that can be exploited, the virome can persist, evade defenses, diversify, and form astonishingly complex and frequently mutualistic, symbiotic interactions with the host. This is possible because of the vast number of lifestyles that can be exploited. The virome can endure because of the interactions between its components. It is necessary to consider that these linkages frequently involve 'trans-kingdom' interactions between different species of viromes and microbiomes that originated in different kingdoms of life (for example, virome interactions with bacteria).

Diversity of Viromes

It is estimated that there are approximately 10^{31} to 10^{32} virus particles in the world at any given time, which is orders of magnitude higher than the number of cells (Mushegian 2020). The biome offers an incredible variety of pathogens, which can infect even the healthiest individuals. The human virome contains antibodies that target bacteria, bacteriophages (phages) that infect archaea, viruses that are infrequently encountered in food, and those capable of infecting human cells (Muhammad et al. 2024c). The human virome consists of these components. Within a single host, many viral subpopulations can coexist. Most studies have been conducted on the microbiota of the human gut, where the most varied communities have been discovered. The human gut is a perfect place for viruses to reproduce, given the abundance of human gut cells and prokaryotic bacteria (Muhammad et al. 2024d). Even while most other parts of a healthy human body have less microbiota, new research shows large viral populations in numerous body parts. Like the reported variation in bacterial and fungal populations, the human virome exhibits substantial inter-individual variations (Table 2). This begs the question: "To what extent do variances in the virome account for observed phenotypic variations?" (Liang and Bushman 2021).

Different characteristics can be used to classify the viruses found in humans. The genetic material that makes up a virus is called its genome, which can be RNA or DNA and double-stranded or single-stranded. The size of a genome can range from just a few kilobases to hundreds of kilobases, depending on the organism (Chaitanya and Chaitanya 2019). The genetic content of every virus is protected inside a protein shell known as a capsid. Viruses may have more than one lipid membrane surrounding them. A virus particle's shape is one factor that can be considered while classifying viruses. Virus particles can take several possible forms, including spherical (typically icosahedral), filamentous, bullet-shaped, pleomorphic, or even tailed, similar to how phages are structured. Phages likely make up most of the human virome (Fermin 2018). In recent years, several studies have been conducted to define the human virome in various body locations, indicating the presence of diverse and prolific populations (Figure 2). Phage populations in different anatomical regions of the body may be very different. This is likely to be the case because the human body is home to a diverse collection of bacteria that serve as hosts. *Microviridae*, which includes phage X174, are icosahedral, non-tail-bearing phages, while the *Caudovirales* are icosahedral, tail-bearing phages. There are also differences in the distribution of eukaryotic viruses at various sites throughout the body. An overview of certain perilous attributes specific to each part is illustrated in Figure 2.

Major Realms of Viruses

The International Council on Viral Taxonomy has recently developed a complete hierarchical taxonomy of viruses; at its highest level, it has

Table 2. Examples of viral population alterations in human disorders.

Human disease	**Sample**	**Major virome alteration**	**References**
Severe acute undernourishment	Feces	Condensed viral variety	(Reyes et al. 2015)
Crohn's disease and ulcerative colitis	Feces	Enlarged *Caudovirales* fertility	(Norman et al. 2015)
Crohn's disease	Feces and biopsies	Reasonable variations	(Pérez-Brocal et al. 2015)
AIDS	Feces	Amplified enteric adenoviruses	(Monaco et al. 2016)
Type 1 diabetes	Feces	Condensed viral diversity	(Zhao et al. 2017a)
Hypertension	Feces	*Erwinia* phage ΦEaH2 and *Lactococcus* phage 1706 may be associated with hypertension.	(Han et al., 2018, Muhammad et al. 2021)
Type 2 diabetes	Feces	Increased putative phage scaffolds	(Ma et al. 2018)
DOCK8 deficiency	Skin swabs	Increased skin virome, especially human papillomavirus	(Tirosh et al. 2018)
Colorectal cancer	Feces	Increased viral diversity	(Nakatsu et al. 2018)
Crohn's disease and ulcerative colitis	Feces	Increased *Caudovirales* abundance	(Fernandes et al. 2019)
Type 1 diabetes during pregnancy	Feces	Amplified picobirnaviruses and tobamoviruses	(Wook Kim et al. 2019)
Bacterial vaginosis	Vaginal swabs	Viral population structures correlated with bacterial vaginosis	(Jakobsen et al. 2020)
Early-diagnosed Crohn's disease and ulcerative colitis	Gut biopsies	Amplified *Hepadnaviridae* and *Hepeviridae*; reduced *Polydnaviridae*, *Tymoviridae* and *Virgaviridae*	(Ungaro et al. 2019)
Coeliac disease autoimmunity	Feces	Increased enteroviruses	(Lindfors et al. 2020)

Crohn's disease	Feces	In their place, milder phages have been introduced.	(Clooney et al. 2019)
Ulcerative colitis	Gut biopsies	Increased virulence of *Caudovirales*, phages, and bacteria, but disappeared connections between viruses and bacteria.	(Zuo et al. 2019)
HIV viraemia	Seminal fluid	Amplified human cytomegalovirus	(Li et al. 2020)
Very early-onset inflammatory bowel disease	Feces	Amplified ratio of *Caudovirales* to *Microviridae*	(Liang et al. 2020)
Haematopoietic stem cell transplantation	Feces	Improved picobirnaviruses	(Legoff et al. 2017, Muhammad et al. 2024b)

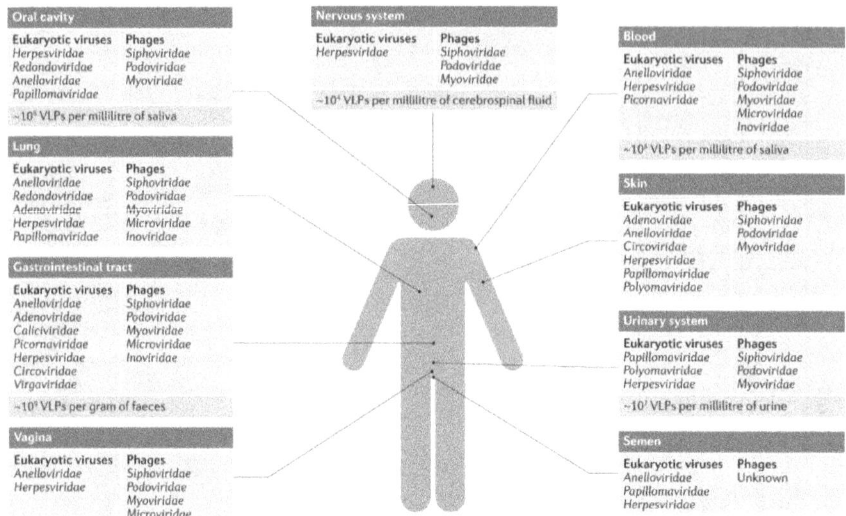

Figure 2. Viruses found at each human body site (Liang and Bushman 2021).

classified six taxa as putatively monophyletic groups (ICTV). The International Committee on Viral Taxonomy (ICTV) created this virus classification system (The new scope of virus taxonomy: partitioning the virosphere into 15 hierarchical ranks 2020). Metaviromics studies on a grand scale have revealed a shocking expansion in the number of virus species present within each of the four major virus domains (Riboviria, Monodnaviria, Duplodnaviria, and Varidnaviria). Viruses are classified as members of the fourth domain, Varidnaviria (Koonin et al. 2020). The environment *Riboviria* includes RNA and DNA viruses capable of reverse transcription and positive-sense, negative-sense, and double-stranded (ds) RNA viruses. Viruses use homologous RNA-dependent RNA polymerases (RdRPs) and reverse transcriptases in this vast universe (RTs). The domain *Monodnaviria* is comprised of circular-genome viruses, most of which are single-stranded DNA (ssDNA) viruses and tiny double-stranded DNA (dsDNA) viruses (papillomaviruses and polyomaviruses). This universe would not hold together without the hallmark gene which encodes a special endonuclease (or a derivative that has been inactivated). This endonuclease is often engaged in the initial rolling circle of replication during genome replication. *Duplodnaviria* is home to dsDNA viruses, including the recently discovered *mirusviruses,* animal herpesviruses, and tailed bacterial and archaeal viruses (Gaïa et al. 2021). The HK97-fold main CP (MCP), the ATPase-nuclease (terminase) responsible for packaging the genome, the portal protein, and the capsid maturation protease are all proteins encoded by the structural gene module of these viruses.

Varidnaviria is the supergroup of viruses that can infect bacteria, archaea, and eukaryotes due to their dsDNA genomes. These pathogens share a common characteristic in their vertical jelly-roll MCPs. The *Bamfordvirae* kingdom contains most viruses because their main capsid proteins fold in a double jelly-roll (DJR) shape.

In contrast, each of the two MCPs seen in Helvetiavirae viruses is composed of a single vertical jelly-roll domain. The International Council for the Taxonomy of Viruses recently acknowledged two new, smaller virus worlds: Adnaviria and Ribozyviria (ICTV). Both rod-shaped and filamentous viruses can be found in the phylum Adnaviria. These viruses infect hyperthermophilic archaea belonging to the phylum Thermoproteota, and they encapsidate linear dsDNA in the A form (Krupovic et al. 2021). In addition to viruses that can infect humans, the domain Ribozyviria also contains viruses that can infect various other animals (Hepojoki et al. 2021). *Ribozyviruses,* like viroids, have nucleocapsid proteins encoded by circular RNA genomes (Delta antigen).

Monophyly is the notion that all viruses can be traced back to a common ancestor. Yet, this is not true for at least three of the four primary realms that exist at present. It remains valid even if we ignore that monophyly depends on almost no VHG. Even though the two kingdoms that make up the realm *Riboviria* have a replicative enzyme identical to the other, it is clear that the origins of these viruses, which emerged from the recruitment of different CPs, were completely separate from one another (Figure 3). It appears that the two groups of Pararnavirae – Ortervirales and Blubervirales – originated from independent retrotransposons, suggesting that *Riboviria* should be separated into at least two new domains to maintain the criterion that taxa should be strictly monophyletic (Gong and Han 2018, Krupovic et al. 2018). The viruses that make up the kingdom Monodnaviria evolved at least four times in a row, each taking a slightly different path. Little rolling-circle plasmids can capture the viral gene that encodes a CP, or cellular genes repurposed as CPs (Kazlauskas et al. 2019). The viruses that make up this region have evolved into what they are now (Figure 3). It was once thought that the two solitary jelly-roll vertical CPs seen in *helvetiaviruases* represented an early version of the MCP in the *Varidnaviria* universe. DJR MCP would have been the result of gene fusion in this scenario. Current research into the cellular origins of the vertical jelly-roll MCPs has revealed that Bamfordvirae and Helvetiavirae form their capsids independently. A recent study (Krupovic et al. 2022) has found that the current state of affairs guarantees the continued existence of the viral planets of *Duplodnaviria, Adnaviria,* and *Ribozyviria,* each with its monophyly. Yet, some viruses do not fit into the already recognized categories; instead, they appear to have evolved independently, making them good candidates for new and more limited types. Archaeal viruses,

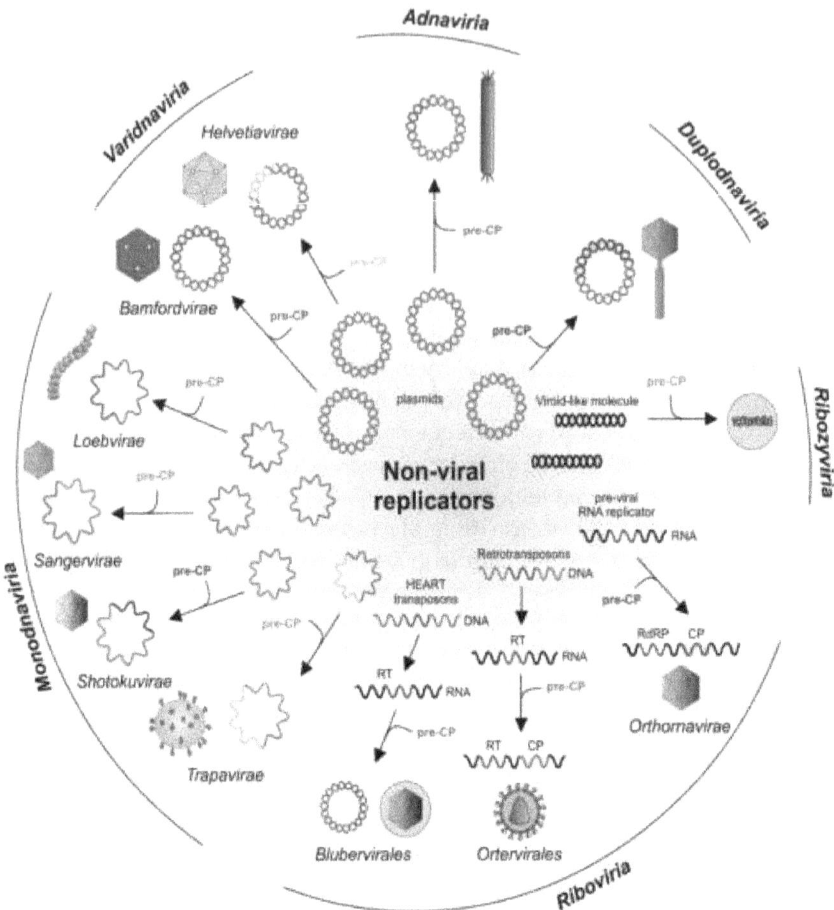

Figure 3. The possible division of virus families and their organization into individual domains would follow taxonomic principles and genetic relationships among these viruses. During the process of virogenesis, non-viral replicators like plasmids and transposons acquire cellular genes that are then repurposed to build viral capsids. This happens because these non-viral replicators have reached cellular DNA. The most monophyletic taxa are emphasized for three of the six categories to illustrate the potential for distinct paths of virogenesis from non-viral replicators. The DNA and RNA replicators, not viruses, are denoted in grey and black, respectively. There are several ancestry branches for capsid proteins, and a distinct color indicates each component. The capsid protein, RNA-dependent RNA polymerase, and reverse transcriptase are each represented by their initials as CP, RdRP, and RT, respectively (Koonin et al. 2023).

giant dsDNA viruses like those in the class Naldaviricetes (bacula-like viruses), and animal *anelloviruses* with small ssDNA genomes appear to be missing the signature rolling-circle endonuclease that binds the members of Monodnaviria together (Taylo et al. 2022). Many virus families fall under the umbrella term 'distinct enigmatic viruses' (Sinha et al. 2023, Muhammad et al. 2024a).

Biofilms

When bacterial colonies cling to a surface and are kept together by a polymer matrix composed of polysaccharides, secreted proteins, and extracellular DNAs, the result is a biofilm. Bacterial colonies form biofilms. Biofilms can colonize various surfaces, including food, medical equipment, and supplies. It is estimated that anywhere from forty to eighty percent of all bacterial cells globally can generate biofilms (Krupovic et al. 2022). In the food industry, forming biofilms by pathogenic bacteria inside processing facilities contributes to the deterioration of food and puts customers at risk. In addition, it has been demonstrated that biofilms can persist on the surfaces of medical devices and in the tissues of hospitalized patients, ultimately resulting in infections that do not go away. As a result of the potentially harmful effects of biofilms on human health and other aspects of society, scientists and members of the general public have devoted significant efforts towards eliminating these substances.

In the seventeenth century, Anton Von Leeuwenhoek was the first to use the term 'biofilm' to describe microbial clumps found on plaque scrapings. Louis Pasteur discovered bacterial exopolysaccharides (EPS) in wine in 1861. EPS is a type of microbial byproduct known as dextran. In 1878, Van Tieghem found the strain of *Leuconostoc mesenteriodes* that produces dextran. These biofilms have existed since 1973, when Characklis first described their indestructibility and disinfectant resistance. Canadian microbiologist William Costerton is credited with inventing the term biofilm in 1978. He originally referred to "an organized community of bacterial cells adhering to an inert or living surface and encased in their polymeric matrix" (Gupta et al. 2019, Yasir et al. 2018). According to the definition provided, a biofilm is "a microbially produced sessile community that is characterized by cells that are irreversibly attached to a substratum or interface or each other" (Donlan and Costerton 2002). According to the authors, these cells "display a changed phenotype concerning growth rate and gene transcription." Also, the authors noted that these cells are "embedded in a matrix of extracellular polymeric molecules that they have generated" (Gupta et al. 2019, Chandki et al. 2011). It has been hypothesized that biofilms play a role in the continued existence of various bacterial diseases. Although there is a chance for favorable outcomes, bacterial biofilms have favorable and unfavorable repercussions (Muhammad et al. 2020). As a result, the production of bacterial biofilm is frequently relevant in agricultural and other industrial settings (Hayta et al. 2021). Current applications for these beneficial biofilms include bioremediation treatment of hazardous pollutants (Sandhya et al. 2022), wastewater treatment (Huang et al. 2019), protection of marine ecosystems (Elnahal et al. 2022), and corrosion prevention (Guo et al. 2022). Only the adverse effects of biofilms on agriculture and industry have been understood for the past several decades. However, biofilms can have both positive and negative impacts on these sectors. As a result, the advantageous components of biofilms have an extraordinary capacity for further development.

Stages of Biofilm Formation

Transforming bacteria from their free-swimming planktonic phase into their biofilm-forming sessile form is a multi-step process that can be considerably complicated. The creation process is influenced by several creative processes, which are regulated by multiple parameters like temperature, pH, gravitational forces, hydrodynamic forces, Brownian motion, the nature of the inhabited surfaces, quorum sensing, secondary messengers, and other signaling molecules. Brownian motion is the random motion that occurs when particles move through a fluid (Rather et al. 2021). Biofilm production can be broken down into four distinct phases (Figure 4).

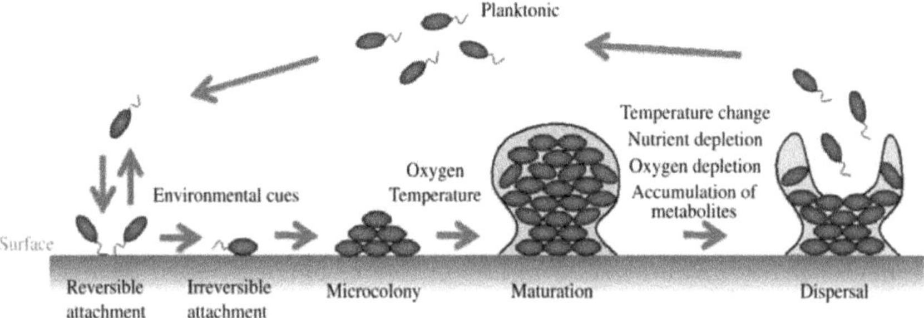

Figure 4. Biofilm development and the influence of specific environmental elements at various times (López and Soto 2019).

Attachment: A Surface-sensing Step

Planktonic microorganisms attaching themselves to surfaces is the first step in developing a biofilm. It is an essential stage in which previously free-floating microorganisms eventually organize themselves into a community structure (Roncoroni et al. 2019). Initial biofilm formation is characterized by the presence of microorganisms attached to surfaces in a manner that is both loose and reversible, as well as by the presence of microorganisms attached to characters in a polarly connected way. Both of these characteristics are present in the initial biofilm formation. After this, the microorganisms settle onto the surfaces, irreversibly affixing themselves and becoming resistant to various physical forces that usually inhibit biofilm formation. In other words, they become immune to biofilm creation (Kimkes and Heinemann 2020). The intracellular signaling molecule known as bis-(3′-5′)-cyclic dimeric guanosine monophosphate (c-di-GMP) plays a vital role in the early phases of biofilm development. It reduces flagella-mediated swimming motility and promotes biofilm matrix creation (Park and Sauer 2022). The Pil-Chp mechanism is a surface-sensing mechanism on the surfaces of microorganisms. Upon attachment, it increases the concentration of c-di-GMP, but once the attachment is severed, the level of this substance returns to its previous level. Therefore, the surface has planktonic cells, which are bacteria that have a low concentration of c-di-GMP and that have not initially encountered characters; they are transformed into surface sentient planktonic cells, which are bacteria that have a high concentration of c-di-GMP and that have discovered surfaces originally, and eventually lead to biofilm formation through the irreversible attachment of cells to characters. In other words, surface nave planktonic cells are transformed into surface sentient planktonic cells (Muhammad et al. 2020).

Growth or Microcolony Formation

When a significant amount of c-di-GMP is present in the surrounding environment, the microorganisms that have successfully established surface connections multiply and aggregate within their own EPS. At some point in time, this led to the establishment of smaller colonies, for contact between germs, surfaces, and cell-cell aggregations that result in microcolonies, flagella, and type IV pili-mediated motility are required for microbial motility (Alotaibi and Bukhari 2021).

Maturation

The extracellular polymeric substance (EPS) helps microbes attach to surfaces, keeps biofilms together, holds clusters cells together, protects against stresses (such as the host's immune response, antibiotics, oxidative damage, and metallic cations), and encases signaling molecules that are essential for quorum sensing (Singh et al. 2021). Because of differences in their metabolic rates and ability to tolerate wind, the bacteria that make up a mature biofilm may cluster into 'mushrooms' or 'towers' (Singh et al. 2022). Through quorum sensing-mediated methods, Otto demonstrated that surfactants and modulins play a part in forming *Staphylococcal* biofilms. A mature biofilm has three distinct layers: the inner regulatory layer, the middle microbial foundation layer, and the outside layer (occupied with planktonic form bacteria ready to exit the biofilm). The control of the central and the outer layers is delegated to the innermost layer (Zhao et al. 2017a).

Dispersion

When the bacteria trapped within a mature biofilm are either actively disseminated (through motility and EPS degradation-dependent dispersion) or passively transferred (by diffusion), a new cycle of biofilm creation begins (due to physical factors such as liquid flow-dependent dispersion). Outgrown populations, intense competition, a lack of nutrients, and enzyme action that causes alginate digestion in *Pseudomonas* spp. Contribute to the dispersal of mature biofilms, as do temperature, oxygen deficiency, and metabolite build-up; up-regulation of genes is responsible for cell motility and EPS degradation, and downregulation of genes is essential for polysaccharide biosynthesis (Rabin et al. 2015, Ajijah et al. 2023).

Microbial Diversity within Biofilms

The microbial variety during biofilm development affects the resultant communities' stability, organization, and implementation. Studies aiming to characterize the microbial diversity inside biofilms, utilizing cutting-edge molecular techniques such as high-throughput sequencing, have increased in recent years. In recent research, up to 2,500 OTUs were discovered at the species level from 16S rRNA gene sequencing in biofilms from drinking water distribution systems (Hüpeden et al. 2020). Hupeden, Wemheuer, Indenbirken, Schulz, and Spieck published these findings. The results of a study on the microbial diversity of biofilms at a wastewater treatment facility revealed a complex web of microbial interactions. Temperature, pH, and the concentration of mixed spirits suspended solids (MLSS) were all found to affect the bacterial activity (Feng et al. 2022). Biofilms have been found in various human body sites, including the mouth, gut, skin, and lungs, and these sites contain a wide variety of microorganisms. These biofilms host extraordinarily varied microbial populations. In addition to archaea, they also consist of viruses, bacteria, and fungi (Curtis et al. 2020). Because the mouth is one of the most densely populated parts of the human body, oral biofilms host various microorganisms (Sedghi et al. 2021). The digestive system is second only to the mouth as the part of the body most populated by bacteria. Gut biofilms host microbial communities that consist of around 770 unique bacterial species. The human mouth is particularly vulnerable to microbial invasion (Kitamoto et al. 2020). The microbial variety of gastrointestinal biofilms has been connected to several diseases and disorders, including obesity, inflammatory bowel disease, and colorectal cancer (Hills Jr et al. 2019). The skin takes up more space than any other organ in the body. Bacterial strains isolated from human skin have been shown to form biofilms in laboratory and animal models. The skin's microbiome is the new focal point of dermatology and cosmetics. It is essential to comprehend the mechanisms responsible for healthy skin and its appearance to keep the skin microbiota, the collection of vital bacteria that reside on our skin, in a state of delicate equilibrium. Dysbiosis, characterized by an imbalance in the skin's microbiota, has been linked to several skin problems. The former group includes conditions like eczema, acne, allergies, and dandruff, while the latter group includes conditions like sensitive skin, irritated skin, and dry skin (Sfriso et al. 2020). Smoking, chronic obstructive pulmonary disease (COPD) and bacterial infections have been found to alter the composition of respiratory biofilms. Periodontitis, asthma, COPD, irritable bowel syndrome and cancer are just some of the disorders that can be exacerbated by the effect of cigarette smoke on the microbiome in the mouth, lungs and digestive system (Huang and Shi 2019).

Application of Biofilms

Given their ubiquitous presence in nature, biofilms are used for a wide range of essential processes, such as bioremediation, wastewater treatment, bio corrosion, and the production of medical devices. Over several decades, industrial operations worldwide have produced waste, leading to sediment contamination and, ultimately, the extensive dispersal of toxins into aquatic environments. More and more ubiquitous contaminants have accumulated and continue to be created and disposed of, increasing the urgency of finding better and more efficient ways to detoxify the environment. Local biological communities have evolved mechanisms for degrading POPs and oxidizing metal contaminants. Yet, they cannot reach the designated clean-up goals because of their low abundance and activity in the ecosystem, difficulties accessing pollution, or environmental constraints. Biofilm communities have many advantages, such as a protective structure, the chance for nutrition and genetic exchange amongst participating microorganisms, and resistance to environmental challenges, including predation and chemicals.

Moreover, biofilm communities allow the exchange of nutrients and genes. In addition, biofilms can be used in many other contexts as biomarkers to track the purity of stream water after activities like mine drainage have contaminated it. These communities are essential to biofilm-mediated remediation solutions and ecosystem monitoring because of their persistence, structure, and diversity of metabolic characteristics (Edwards and Kjellerup 2013).

Implant-related infections and low osteogenic activity in patients can play a role in implant failure or prolonged recovery. The financial and physical tolls are unavoidable (Min et al. 2016). There has been a lot of curiosity about treating diseases by manipulating bacteria. The hypoxic microenvironment of tumor tissues is a crucial element to be considered while treating cancer, and certain bacteria, such as *Salmonella, Escherichia,* and *Listeria,* can come into play in such conditions (Chowdhury et al. 2019). Non-pathogenic bacteria can suppress tumor growth through genetic engineering or mixing them with nanomaterials (Zheng et al. 2018). Probiotics may aid gastrointestinal health by contributing to a balanced microbiota in the intestines. Probiotics, which are non-pathogenic organisms, are utilized for the treatment of a wide variety of digestive ailments like infectious diarrhea, inflammatory bowel disease, and irritable bowel syndrome due to their ability to eradicate pathogenic bacteria and control the immune system of the host (Sharma and Sharma 2021). In a new study, researchers modified the surface of titanium implants using zymosan, a fungal polysaccharide, to increase the likelihood of successful implant-bone integration. Improved implant-bone integration was achieved by stimulating macrophages to produce osteogenic cytokines (Shi et al. 2018). It is well known that bacteria and the EPS matrix surrounding the bacteria make up bacterial biofilms. It is also well known that *L. casei* has polysaccharides in its cell walls and extracellular polymeric substance (EPS, made up of polysaccharides, proteins and nucleic acids). Hence, the exposed polysaccharides of *L. casei* biofilms readily trigger the activity of macrophages upon contact with *L. casei*. In order to prevent non-inactivated bacteria from colonising the Ti substrates, biofilms of L. casei are cultivated on their surface and then rendered inactive through exposure to ultraviolet (UV) irradiation. The inactivated *L. casei* biofilm hinders MRSA infection and enhances the healing of bone tissue. Given that the active component of *L. casei* biofilms can endure short-term exposure to UV irradiation (Tan et al. 2020).

Conclusion

Microbial ecology is an exciting and rapidly developing field that has important implications for our understanding of the natural world and the development of novel applications in agriculture, health, biotechnology, and environmental studies. Microbiomes, viromes, and biofilms—the three main components of microbial ecology—have all seen increased research funding and attention. Viromes and microbiomes share an ecological niche; biofilms are populations of bacteria that cling to a surface and form distinct structures with their traits and activities. Scientists are interested in many of these aspects of microbial ecology because of their biotechnological potential and usefulness in defining ecosystem features and functions. The most important thing we have learned from microbial ecology is that microbial communities are diverse in taxonomy and position. Diversity may help microbial communities adapt and survive.

Interactions between microbial species affect the nutrient cycle, ecological stability, and disease susceptibility, making microbial ecology significant. Microbial ecology research has produced several breakthroughs and technologies with practical uses. Due to their metabolism, biofilm communities are used in bioremediation and wastewater treatment. Microbiome analysis can also diagnose and treat inflammatory bowel disease and microbial infections, enabling precision medicine and tailored therapeutics. These advances require further research and awareness of microbial ecology and its uses. We must collaborate across academic fields to understand the principles that manage microbial populations and apply this knowledge to the earth's problems. To improve microbial community investigations, high-throughput sequencing and bioinformatics technologies are proposed. To understand the resilience and flexibility of microbial communities, further studies are required to elucidate the mechanisms underpinning microbial species interactions. We must encourage cross-disciplinary collaboration and the application of microbial ecology research to real-world problems. Lastly, microbial ecology needs more public engagement and education to help people understand how bacteria shape our world and how we might use them.

Acknowledgments

This work was financially supported by the Third Xinjiang Scientific Expedition Program (Grant No. 2022xjkk1200), the National Natural Science Foundation of China (Nos: 32061143043 and 31972856), the Key-Area Research and Development Program of Guangdong Province (No. 2022B0202110001), and the Xinjiang Uygur Autonomous Region regional coordinated innovation project (Shanghai cooperation organization science and technology partnership program) (No. 2021E01018).

References

Abdelghafar, A., Yousef, N. and Askoura, M. (2022). Zinc oxide nanoparticles reduce biofilm formation, synergize antibiotics action and attenuate Staphylococcus aureus virulence in host; an important message to clinicians. BMC Microbiology, 22: 1–17.

Agans, R., Gordon, A., Kramer, D. L., Perez-Burillo, S., Rufián-Henares, J. A. and Paliy, O. (2018). Dietary fatty acids sustain the growth of the human gut microbiota. Applied and Environmental Microbiology, 84: e01525-18.

Ajijah, N., Fiodor, A., Pandey, A. K., Rana, A. and Pranaw, K. (2023). Plant Growth-Promoting Bacteria (PGPB) with biofilm-forming ability: A multifaceted agent for sustainable agriculture. Diversity, 15: 112.

Alotaibi, G. F. and Bukhari, M. A. (2021). Factors influencing bacterial biofilm formation and development. Am. J. Biomed. Sci. Res., 12: 617–626.

Amaroli, A., Ravera, S., Zekiy, A., Benedicenti, S. and Pasquale, C. (2022). A narrative review on oral and periodontal bacteria Microbiota photobiomodulation, through visible and near-infrared light: From the origins to modern therapies. International Journal of Molecular Sciences, 23: 1372.

Balan, P., Seneviratne, C. J. and Crielaard, W. (2017). Composition and Diversity of Human Oral Microbiome. Microbial Biofilms. CRC Press.

Bardgett, R. D. and Wardle, D. A. (2010). Aboveground-belowground Linkages: Biotic Interactions, Ecosystem Processes, and Global Change. Oxford University Press.

Barko, P., Mcmichael, M., Swanson, K. S. and Williams, D. A. (2018). The gastrointestinal microbiome: A review. Journal of Veterinary Internal Medicine, 32: 9–25.

Berhe, N., Tefera, Y. and Tintagu, T. (2017). Review on biofilm formation and its control options. International Journal of Advanced Research in Biological Sciences, 8: 122–133.

Bowler, P. G. (2018). Antibiotic resistance and biofilm tolerance: A combined threat in the treatment of chronic infections. Journal of Wound Care, 27: 273–277.

Boxberger, M., Cenizo, V., Cassir, N. and La Scola, B. (2021). Challenges in exploring and manipulating the human skin microbiome. Microbiome, 9: 1–14.

Byrd, A. L., Belkaid, Y. and Segre, J. A. (2018). The human skin microbiome. Nature Reviews Microbiology, 16: 143–155.

Cadwell, K. (2015). The virome in host health and disease. Immunity, 42: 805–813.

Chaitanya, K. and Chaitanya, K. (2019). Structure and organization of virus genomes. Genome and Genomics: From Archaea to Eukaryotes, 1–30.

Chandki, R., Banthia, P. and Banthia, R. (2011). Biofilms: A microbial home. Journal of Indian Society of Periodontology, 15: 111.

Chevallereau, A., Pons, B. J., Van Houte, S. and Westra, E. R. (2022). Interactions between bacterial and phage communities in natural environments. Nature Reviews Microbiology, 20: 49–62.

Chowdhury, S., Castro, S., Coker, C., Hinchliffe, T. E., Arpaia, N. and Danino, T. (2019). Programmable bacteria induce durable tumor regression and systemic antitumor immunity. Nature Medicine, 25: 1057–1063.

Clooney, A. G., Sutton, T. D., Shkoporov, A. N., Holohan, R. K., Daly, K. M., O'regan, O., Ryan, F. J., Draper, L. A., Plevy, S. E. and Ross, R. P. (2019). Whole-virome analysis sheds light on viral dark matter in inflammatory bowel disease. Cell Host & Microbe, 26: 764–778. e5.

Curtis, M. A., Diaz, P. I. and Van Dyke, T. E. (2020). The role of the microbiota in periodontal disease. Periodontology, 83: 14–25.

Davis, C. D. (2016). The gut microbiome and its role in obesity. Nutrition Today, 51: 167.

Decker, A. M., Kapila, Y. L. and Wang, H. L. (2021). The psychobiological links between chronic stress-related diseases, periodontal/peri-implant diseases, and wound healing. Periodontology, 87: 94–106.

Deo, P. N. and Deshmukh, R. (2019). Oral microbiome: Unveiling the fundamentals. Journal of Oral and Maxillofacial Pathology: JOMFP, 23: 122.

Dewasthale, S., Mani, I. and Vasdev, K. (2018). Microbial biofilm: Current challenges in health care industry. J. Appl. Biotechnol. Bioeng., 5.

Edwards, S. J. and Kjellerup, B. V. (2013). Applications of biofilms in bioremediation and biotransformation of persistent organic pollutants, pharmaceuticals/personal care products, and heavy metals. Applied Microbiology and Biotechnology, 97: 9909–9921.

Elnahal, A. S., El-Saadony, M. T., Saad, A. M., Desoky, E. -S. M., El-Tahan, A. M., Rady, M. M., Abuqamar, S. F. and El-Tarabily, K. A. (2022). The use of microbial inoculants for biological control, plant growth promotion, and sustainable agriculture: A review. European Journal of Plant Pathology, 162: 759–792.

Esposito, A. M., Esposito, M. M. and Ptashnik, A. (2022). Phylogenetic diversity of animal oral and gastrointestinal viromes useful in surveillance of zoonoses. Microorganisms, 10: 1815.

Falcao, L. and Inaturals, S. (2021). The crosstalk between skin microbiome and skin immunity. Health.

Feng, Z., Li, T., Lin, Y. and Wu, G. (2022). Microbial communities and interactions in full-scale A2/O and MBR wastewater treatment plants. Journal of Water Process Engineering, 46: 102660.

Fermin, G. (2018). Virion structure, genome organization, and taxonomy of viruses. Viruses, 17.

Fernandes, M. A., Verstraete, S. G., Phan, T. G., Deng, X., Stekol, E., Lamere, B., Lynch, S. V., Heyman, M. B. and Delwart, E. (2019). Enteric virome and bacterial microbiota in children with ulcerative colitis and Crohn's disease. Journal of Pediatric Gastroenterology and Nutrition, 68: 30.

Foo, J. L., Ling, H., Lee, Y. S. and Chang, M. W. (2017). Microbiome engineering: Current applications and its future. Biotechnology Journal, 12: 1600099.

Gaïa, M., Meng, L., Pelletier, E., Forterre, P., Vanni, C., Fernandez-Guerra, A., Jaillon, O., Wincker, P., Ogata, H. and Krupovic, M. (2021). Plankton-infecting relatives of herpesviruses clarify the evolutionary trajectory of giant viruses. bioRxiv, 2021.12. 27.474232.

Gomaa, E. Z. (2020). Human gut microbiota/microbiome in health and diseases: A review. Antonie Van Leeuwenhoek, 113: 2019–2040.

Gong, Z. and Han, G. -Z. (2018). Insect retroelements provide novel insights into the origin of hepatitis B viruses. Molecular Biology and Evolution, 35: 2254–2259.

Guo, N., Zhao, Q., Hui, X., Guo, Z., Dong, Y., Yin, Y., Zeng, Z. and Liu, T. (2022). Enhanced corrosion protection action of biofilms based on endogenous and exogenous bacterial cellulose. Corrosion Science, 194: 109931.

Gupta, P., Pruthi, P. A. and Pruthi, V. (2019). Role of exopolysaccharides in biofilm formation. Introduction to biofilm engineering. ACS Publications.

Han, M., Yang, P., Zhong, C. and Ning, K. (2018). The human gut virome in hypertension. Frontiers in Microbiology, 9: 3150.

Hayta, E. N., Ertelt, M. J., Kretschmer, M. and Lieleg, O. (2021). Bacterial materials: Applications of natural and modified biofilms. Advanced Materials Interfaces, 8: 2101024.

Hepojoki, J., Hetzel, U., Paraskevopoulou, S., Drosten, C., Harrach, B., Zerbini, M., Koonin, E., Krupovic, M., Dolja, V. and Kuhn, J. (2021). Create one new realm (Ribozyviria) including one new family (Kolmioviridae) including genus Deltavirus and seven new genera for a total of 15 species. ResearchGate doi: https://doi. org/10.13140/RG, 2.

Hidalgo-Cantabrana, C., Sanozky-Dawes, R. and Barrangou, R. (2018). Insights into the human virome using CRISPR spacers from microbiomes. Viruses, 10: 479.

Hills Jr, R. D., Pontefract, B. A., Mishcon, H. R., Black, C. A., Sutton, S. C. and Theberge, C. R. (2019). Gut microbiome: profound implications for diet and disease. Nutrients, 11: 1613.

Huang, C. and Shi, G. (2019). Smoking and microbiome in oral, airway, gut and some systemic diseases. Journal of Translational Medicine, 17: 1–15.

Huang, H., Peng, C., Peng, P., Lin, Y., Zhang, X. and Ren, H. (2019). Towards the biofilm characterization and regulation in biological wastewater treatment. Applied Microbiology and Biotechnology, 103: 1115–1129.

Hüpeden, J., Wemheuer, B., Indenbirken, D., Schulz, C. and Spieck, E. (2020). Taxonomic and functional profiling of nitrifying biofilms in freshwater, brackish and marine RAS biofilters. Aquacultural Engineering, 90: 102094.

Idris, A., Hasnain, S. Z., Huat, L. Z. and Koh, D. (2017). Human diseases, immunity and the oral microbiota—Insights gained from metagenomic studies. Oral Science International, 14: 27–32.

Jakobsen, R. R., Haahr, T., Humaidan, P., Jensen, J. S., Kot, W. P., Castro-Mejia, J. L., Deng, L., Leser, T. D. and Nielsen, D. S. (2020). Characterization of the vaginal DNA virome in health and dysbiosis. Viruses, 12: 1143.

Jenkins, T. P. (2019). Exploring the impact of gastrointestinal parasitic helminths on the human microbiome using advanced biomolecular and bioinformatics technologies. University of Cambridge.

Kazlauskas, D., Varsani, A., Koonin, E. V. and Krupovic, M. (2019). Multiple origins of prokaryotic and eukaryotic single-stranded DNA viruses from bacterial and archaeal plasmids. Nature Communications, 10: 3425.

Kim, B. -R., Shin, J., Guevarra, R. B., Lee, J. H., Kim, D. W., Seol, K. -H., Lee, J. -H., Kim, H. B. and Isaacson, R. E. (2017). Deciphering diversity indices for a better understanding of microbial communities. Journal of Microbiology and Biotechnology, 27: 2089–2093.

Kimkes, T. E. and Heinemann, M. (2020). How bacteria recognise and respond to surface contact. FEMS Microbiology Reviews, 44: 106–122.

Kitamoto, S., Nagao-Kitamoto, H., Hein, R., Schmidt, T. and Kamada, N. (2020). The bacterial connection between the oral cavity and the gut diseases. Journal of Dental Research, 99: 1021–1029.

Kogut, M. H., Lee, A. and Santin, E. (2020). Microbiome and pathogen interaction with the immune system. Poultry Science, 99: 1906–1913.

Kokare, C., Chakraborty, S., Khopade, A. and Mahadik, K. R. (2009). Biofilm: Importance and applications.

Koonin, E. V., Dolja, V. V., Krupovic, M., Varsani, A., Wolf, Y. I., Yutin, N., Zerbini, F. M. and Kuhn, J. H. (2020). Global organization and proposed megataxonomy of the virus world. Microbiology and Molecular Biology Reviews, 84: e00061–19.

Koonin, E. V., Krupovic, M. and Dolja, V. V. (2023). The global virome: How much diversity and how many independent origins? Wiley Online Library.

Krupovic, M., Blomberg, J., Coffin, J. M., Dasgupta, I., Fan, H., Geering, A. D., Gifford, R., Harrach, B., Hull, R. and Johnson, W. (2018). Ortervirales: New virus order unifying five families of reverse-transcribing viruses. Journal of Virology, 92: e00515–18.

Krupovic, M., Kuhn, J. H., Wang, F., Baquero, D. P., Dolja, V. V., Egelman, E. H., Prangishvili, D. and Koonin, E. V. (2021). Adnaviria: A new realm for archaeal filamentous viruses with linear A-form double-stranded DNA genomes. Journal of Virology, 95: e00673–21.

Krupovic, M., Makarova, K. S. and Koonin, E. V. (2022). Cellular homologs of the double jelly-roll major capsid proteins clarify the origins of an ancient virus kingdom. Proceedings of the National Academy of Sciences, 119: e2120620119.

Kundu Smita, S. and Rana, D. G. (2021). Therapeutic role and importance of gut microbiota in brain and related disorder: A. Asian Journal of Pharmacy and Pharmacology, 7: 47–57.

Legoff, J., Michonneau, D. and Socie, G. (2020). The virome in hematology—Stem cell transplantation and beyond. Seminars in Hematology. Elsevier, 19–25.

Legoff, J., Resche-Rigon, M., Bouquet, J., Robin, M., Naccache, S. N., Mercier-Delarue, S., Federman, S., Samayoa, E., Rousseau, C. and Piron, P. (2017). The eukaryotic gut virome in hematopoietic stem cell transplantation: New clues in enteric graft-versus-host disease. Nature Medicine, 23: 1080–1085.

Li, X. -H. and Lee, J. -H. (2017). Antibiofilm agents: A new perspective for antimicrobial strategy. Journal of Microbiology, 55: 753–766.

Li, Y., Altan, E., Pilcher, C., Hartogensis, W., Hecht, F. M., Deng, X. and Delwart, E. (2020). Semen virome of men with HIV on or off antiretroviral treatment. AIDS, 34: 827–832.

Liang, G. and Bushman, F. D. (2021). The human virome: Assembly, composition and host interactions. Nature Reviews Microbiology, 19: 514–527.

Liang, G., Conrad, M. A., Kelsen, J. R., Kessler, L. R., Breton, J., Albenberg, L. G., Marakos, S., Galgano, A., Devas, N. and Erlichman, J. (2020). Dynamics of the stool virome in very early-onset inflammatory bowel disease. Journal of Crohn's and Colitis, 14: 1600–1610.

Lindfors, K., Lin, J., Lee, H. -S., Hyöty, H., Nykter, M., Kurppa, K., Liu, E., Koletzko, S., Rewers, M. and Hagopian, W. (2020). Metagenomics of the faecal virome indicate a cumulative effect of enterovirus and gluten amount on the risk of coeliac disease autoimmunity in genetically at risk children: the TEDDY study. Gut, 69: 1416–1422.

Litchman, E. (2010). Invisible invaders: Non-pathogenic invasive microbes in aquatic and terrestrial ecosystems. Ecology Letters, 13: 1560–1572.

Lloyd-Price, J., Abu-Ali, G. and Huttenhower, C. (2016). The healthy human microbiome. Genome Medicine, 8: 1–11.

López, Y. and Soto, S. M. (2019). The usefulness of microalgae compounds for preventing biofilm infections. Antibiotics, 9: 9.

Luo, J., Lin, X., Bordiga, M., Brennan, C. and Xu, B. (2021). Manipulating effects of fruits and vegetables on gut microbiota—A critical review. International Journal of Food Science & Technology, 56: 2055–2067.

Ma, Y., You, X., Mai, G., Tokuyasu, T. and Liu, C. (2018). A human gut phage catalog correlates the gut phageome with type 2 diabetes. Microbiome, 6: 1–12.

Makki, K., Deehan, E. C., Walter, J. and Bäckhed, F. (2018). The impact of dietary fiber on gut microbiota in host health and disease. Cell Host & Microbe, 23: 705–715.

Malla, M. A., Dubey, A., Kumar, A., Yadav, S., Hashem, A. and Abd_allah, E. F. (2019). Exploring the human microbiome: the potential future role of next-generation sequencing in disease diagnosis and treatment. Frontiers in Immunology, 9: 2868.

Mcloughlin, I. J., Wright, E. M., Tagg, J. R., Jain, R. and Hale, J. D. (2022). Skin microbiome—the next frontier for probiotic intervention. Probiotics and Antimicrobial Proteins, 14: 630–647.

Min, J., Choi, K. Y., Dreaden, E. C., Padera, R. F., Braatz, R. D., Spector, M. and Hammond, P. T. (2016). Designer dual therapy nanolayered implant coatings eradicate biofilms and accelerate bone tissue repair. ACS Nano, 10: 4441–4450.

Monaco, C. L., Gootenberg, D. B., Zhao, G., Handley, S. A., Ghebremichael, M. S., Lim, E. S., Lankowski, A., Baldridge, M. T., Wilen, C. B. and Flagg, M. (2016). Altered virome and bacterial microbiome in human immunodeficiency virus-associated acquired immunodeficiency syndrome. Cell Host & Microbe, 19: 311–322.

Mosaddad, S. A., Tahmasebi, E., Yazdanian, A., Rezvani, M. B., Seifalian, A., Yazdanian, M. and Tebyanian, H. (2019). Oral microbial biofilms: An update. European Journal of Clinical Microbiology & Infectious Diseases, 38: 2005–2019.

Muhammad, M., Ahmad, J., Basit, A., Mohamed, H. I., Khan, A. and Kamel, E. A. (2024a). Antimicrobial activity of Penicillium species metabolites. Fungal Secondary Metabolites. Elsevier.

Muhammad, M., Ahmad, M. W., Basit, A., Ullah, S., Mohamed, H. I., Nisar, N. and Khan, A. (2024b). Plant growth-promoting rhizobacteria and their applications and role in the management of soilborne diseases. Bacterial Secondary Metabolites. Elsevier.

Muhammad, M., Badshah, L., Shah, A. A., Shah, M. A., Abdullah, A., Bussmann, R. W. and Basit, A. (2021). Ethnobotanical profile of some useful plants and fungi of district Dir Upper, Tehsil Darora, Khyber Pakhtunkhwa, Pakistan. Ethnobotany Research and Applications, 21: 1–15.

Muhammad, M., Basit, A., Majeed, M., Shah, A. A., Ullah, I., Mohamed, H. I., Khan, A. and Ghanaim, A. M. (2024c). Bacterial pigments and their applications. Bacterial Secondary Metabolites. Elsevier.

Muhammad, M., Begum, S., Basit, A., Arooj, A. and Mohamed, H. I. (2024d). Bacterial enzymes and their application in agroecology. Bacterial Secondary Metabolites. Elsevier.

Muhammad, M. H., Idris, A. L., Fan, X., Guo, Y., Yu, Y., Jin, X., Qiu, J., Guan, X. and Huang, T. (2020). Beyond risk: Bacterial biofilms and their regulating approaches. Frontiers in Microbiology, 11: 928.

Mushegian, A. (2020). Are there 1031 virus particles on earth, or more, or fewer? Journal of Bacteriology, 202: e00052–20.

Nakatsu, G., Zhou, H., Wu, W. K. K., Wong, S. H., Coker, O. O., Dai, Z., Li, X., Szeto, C. -H., Sugimura, N. and Lam, T. Y. -T. (2018). Alterations in enteric virome are associated with colorectal cancer and survival outcomes. Gastroenterology, 155: 529–541. e5.

Nguyen, U. T. and Kalan, L. R. (2022). Forgotten fungi: The importance of the skin mycobiome. Current Opinion in Microbiology, 70: 102235.

Norman, J. M., Handley, S. A., Baldridge, M. T., Droit, L., Liu, C. Y., Keller, B. C., Kambal, A., Monaco, C. L., Zhao, G. and Fleshner, P. (2015). Disease-specific alterations in the enteric virome in inflammatory bowel disease. Cell, 160: 447–460.

Park, S. and Sauer, K. (2022). Controlling biofilm development through cyclic di-GMP signaling. Pseudomonas aeruginosa: Biology, Pathogenesis and Control Strategies, 69–94.

Pérez-Brocal, V., García-López, R., Nos, P., Beltrán, B., Moret, I. and Moya, A. (2015). Metagenomic analysis of Crohn's disease patients identifies changes in the virome and microbiome related to disease status and therapy, and detects potential interactions and biomarkers. Inflammatory Bowel Diseases, 21: 2515–2532.

Porras-Alfaro, A. and Bayman, P. (2011). Hidden fungi, emergent properties: Endophytes and microbiomes. Annual Review of Phytopathology, 49: 291–315.

Rabin, N., Zheng, Y., Opoku-Temeng, C., Du, Y., Bonsu, E. and Sintim, H. O. (2015). Biofilm formation mechanisms and targets for developing antibiofilm agents. Future Medicinal Chemistry, 7: 493–512.

Rather, M. A., Gupta, K. and Mandal, M. (2021). Microbial biofilm: Formation, architecture, antibiotic resistance, and control strategies. Brazilian Journal of Microbiology, 1–18.

Reyes, A., Blanton, L. V., Cao, S., Zhao, G., Manary, M., Trehan, I., Smith, M. I., Wang, D., Virgin, H. W. and Rohwer, F. (2015). Gut DNA viromes of Malawian twins discordant for severe acute malnutrition. Proceedings of the National Academy of Sciences, 112: 11941–11946.

Rinninella, E., Raoul, P., Cintoni, M., Franceschi, F., Miggiano, G. A. D., Gasbarrini, A. and Mele, M. C. (2019). What is the healthy gut microbiota composition? A changing ecosystem across age, environment, diet, and diseases. Microorganisms, 7: 14.

Roncoroni, M., Brandani, J., Battin, T. I. and Lane, S. N. (2019). Ecosystem engineers: Biofilms and the ontogeny of glacier floodplain ecosystems. Wiley Interdisciplinary Reviews: Water, 6: e1390.

Sandhya, M., Huang, Y., Li, J., Wu, X., Zhou, Z., Lei, Q., Bhatt, P. and Chen, S. (2022). Biofilm-mediated bioremediation is a powerful tool for the removal of environmental pollutants. Chemosphere, 133609.

Santoro, A., Tomino, C., Prinzi, G., Cardaci, V., Fini, M., Macera, L., Russo, P. and Maggi, F. (2020). Microbiome in chronic obstructive pulmonary disease: Role of natural products against microbial pathogens. Current Medicinal Chemistry, 27: 2931–2948.

Sedghi, L., Dimassa, V., Harrington, A., Lynch, S. V. and Kapila, Y. L. (2021). The oral microbiome: Role of key organisms and complex networks in oral health and disease. Periodontology, 87: 107–131.

Sfriso, R., Egert, M., Gempeler, M., Voegeli, R. and Campiche, R. (2020). Revealing the secret life of skin-with the microbiome you never walk alone. International Journal of Cosmetic Science, 42: 116–126.

Shade, A., Peter, H., Allison, S. D., Baho, D. L., Berga, M., Bürgmann, H., Huber, D. H., Langenheder, S., Lennon, J. T. and Martiny, J. B. (2012). Fundamentals of microbial community resistance and resilience. Frontiers in Microbiology, 3: 417.

Sharma, N., Bhatia, S., Sodhi, A. S. and Batra, N. (2018). Oral microbiome and health. AIMS Microbiology, 4: 42.

Sharma, V. and Sharma, P. (2021). Probiotics as anti-inflammatory agents in inflammatory bowel disease and irritable bowel syndrome. Probiotic Research in Therapeutics: Volume 2: Modulation of Gut Flora: Management of Inflammation and Infection Related Gut Etiology, 123–139.

Shi, Y., Wang, L., Niu, Y., Yu, N., Xing, P., Dong, L. and Wang, C. (2018). Fungal component coating enhances titanium implant-bone integration. Advanced Functional Materials, 28: 1804483.

Singh, A., Padmesh, S., Dwivedi, M. and Kostova, I. (2022). How good are bacteriophages as an alternative therapy to mitigate biofilms of nosocomial infections. Infection and Drug Resistance, 503–532.

Singh, S., Datta, S., Narayanan, K. B. and Rajnish, K. N. (2021). Bacterial exo-polysaccharides in biofilms: Role in antimicrobial resistance and treatments. Journal of Genetic Engineering and Biotechnology, 19: 1–19.

Sinha, D., Odoh, U. E., Ganguly, S., Muhammad, M., Chatterjee, M., Chikeokwu, I. and Egbuna, C. (2023). Phytochemistry, history, and progress in drug discovery. Phytochemistry, Computational Tools and Databases in Drug Discovery. Elsevier.

Smith, L. K. and Wissel, E. F. (2019). Microbes and the mind: How bacteria shape affect, neurological processes, cognition, social relationships, development, and pathology. Perspectives on Psychological Science, 14: 397–418.

Sousa, R., Morais, P., Dias, E. and Antunes, C. (2011). Biological invasions and ecosystem functioning: Time to merge. Biological Invasions, 13: 1055–1058.

Sultan, A. S., Kong, E. F., Rizk, A. M. and Jabra-Rizk, M. A. (2018). The oral microbiome: A Lesson in coexistence. PLoS Pathogens, 14: e1006719.

Tan, L., Fu, J., Feng, F., Liu, X., Cui, Z., Li, B., Han, Y., Zheng, Y., Yeung, K. W. K. and Li, Z. (2020). Engineered probiotics biofilm enhances osseointegration via immunoregulation and anti-infection. Science Advances, 6: eaba5723.

Taylo, L. J., Keeler, E. L., Bushman, F. D. and Collman, R. G. (2022). The enigmatic roles of Anelloviridae and Redondoviridae in humans. Current Opinion in Virology, 55: 101248.

Thursby, E. and Juge, N. (2017). Introduction to the human gut microbiota. Biochemical Journal, 474: 1823–1836.

Tirosh, O., Conlan, S., Deming, C., Lee-Lin, S. -Q., Huang, X., Su, H. C., Freeman, A. F., Segre, J. A. and Kong, H. H. (2018). Expanded skin virome in DOCK8-deficient patients. Nature Medicine, 24: 1815–1821.

Umar, Z., Umar, A., Abdulsalam, M., Vantsawa, P., Idris, H., Adewole, A., Suleiman, A. and Abubakar, A. (2022). Influence of feeding habit and antibiotic use on distribution of bifidobacteria in saliva of infants in kaduna metropolis. Biosciences Journal of Fudma, 3: 7–14.

Ungaro, F., Massimino, L., Furfaro, F., Rimoldi, V., Peyrin-Biroulet, L., D'alessio, S. and Danese, S. (2019). Metagenomic analysis of intestinal mucosa revealed a specific eukaryotic gut virome signature in early-diagnosed inflammatory bowel disease. Gut Microbes, 10: 149–158.

Van Der Putten, W. H., Klironomos, J. N. and Wardle, D. A. (2007). Microbial ecology of biological invasions. The ISME Journal, 1: 28–37.

Wiley, N., Dinan, T., Ross, R., Stanton, C., Clarke, G. and Cryan, J. (2017). The microbiota-gut-brain axis as a key regulator of neural function and the stress response: Implications for human and animal health. Journal of Animal Science, 95: 3225–3246.

Wook Kim, K., Allen, D. W., Briese, T., Couper, J. J., Barry, S. C., Colman, P. G., Cotterill, A. M., Davis, E. A., Giles, L. C. and Harrison, L. C. (2019). Distinct gut virome profile of pregnant women with type 1 diabetes in the ENDIA study. Open forum infectious diseases. Oxford University Press US, ofz025.

Yao, S., Hao, L., Zhou, R., Jin, Y., Huang, J. and Wu, C. (2022). Multispecies biofilms in fermentation: Biofilm formation, microbial interactions, and communication. Comprehensive Reviews in Food Science and Food Safety, 21: 3346–3375.

Yasir, M., Willcox, M. D. P. and Dutta, D. (2018). Action of antimicrobial peptides against bacterial biofilms. Materials, 11: 2468.

Zárate, S., Taboada, B., Yocupicio-Monroy, M. and Arias, C. F. (2017). Human virome. Archives of Medical Research, 48: 701–716.

Zhao, G., Droit, L., Gilbert, M. H., Schiro, F. R., Didier, P. J., Si, X., Paredes, A., Handley, S. A., Virgin, H. W. and Bohm, R. P. (2019). Virome biogeography in the lower gastrointestinal tract of rhesus macaques with chronic diarrhea. Virology, 527: 77–88.

Zhao, G., Vatanen, T., Droit, L., Park, A., Kostic, A. D., Poon, T. W., Vlamakis, H., Siljander, H., Härkönen, T. and Hämäläinen, A. -M. (2017a). Intestinal virome changes precede autoimmunity in type I diabetes-susceptible children. Proceedings of the National Academy of Sciences, 114: E6166–E6175.

Zhao, H., Chu, M., Huang, Z., Yang, X., Ran, S., Hu, B., Zhang, C. and Liang, J. (2017b). Variations in oral microbiota associated with oral cancer. Scientific Reports, 7: 11773.

Zheng, D. -W., Chen, Y., Li, Z. -H., Xu, L., Li, C. -X., Li, B., Fan, J. -X., Cheng, S. -X. and Zhang, X. -Z. (2018). Optically-controlled bacterial metabolite for cancer therapy. Nature Communications, 9: 1680.

Zilber-Rosenberg, I. and Rosenberg, E. (2021). Microbial-driven genetic variation in holobionts. FEMS Microbiology Reviews, 45: fuab022.

Zuo, T., Lu, X. -J., Zhang, Y., Cheung, C. P., Lam, S., Zhang, F., Tang, W., Ching, J. Y., Zhao, R. and Chan, P. K. (2019). Gut mucosal virome alterations in ulcerative colitis. Gut, 68: 1169–1179.

CHAPTER 2

Microbiomes and Viromes in Infection

Shanmuga Sundar S.,[1,*] *Kannan N.*,[2] *Devika R.*,[1] *Karma Gyurmey Dolma*[3] and *Veeranoot Nissapatorn*[4,5]

Introduction

Microbiomes are collections of microorganisms that occupy various regions of the human body and other living species, including bacteria, fungi, viruses and other microbes. These microorganisms reside in a complex ecosystem and serve vital roles in human health by regulating the immune system, boosting nutrient absorption and preventing harmful bacteria from colonising. There are various microbiomes in the human body, including the gut, skin, mouth and vaginal microbiomes. Each of these microbiomes is made up of a distinct mix of bacteria that perform specialised roles. In recent decades, exploration of the gut microbiome has advanced rapidly towards sequence-based screening and a gnotobiotic

[1] Department of Biotechnology, Aarupadai Veedu Institute of Technology, Vinayaka Mission's Research Foundation (DU), Paiyanoor – 603 104, Chennai, Tamil Nadu, India.

[2] Department of Biotechnology, Manipal Institute of Technology, Manipal Academy of Higher Education, Manipal, Karnataka-576104.

[3] Department of Microbiology, Sikkim Manipal University, India.

[4] Department of Medical Technology, School of Allied Health Sciences, Walailak University, Nakhon Si Thammarat, 80160, Thailand.

[5] School of Allied Health Sciences, Southeast Asia Water Team (SEA Water Team) and World Union for Herbal Drug Discovery (WUHeDD), and Research Excellence Centerfor Innovation and Health Products, Walailak University, Nakhon Si Thammarat, 80160, Thailand.

* Corresponding author: sundarannauniv85@gmail.com

model in humans (Kho and Lal 2018). The role of the gut microbial population is unique in terms of host physiological functions, homeostasis of immunity, metabolism, dysfunction in the nervous system, etc. (Braat et al. 2006). The human microbiota varies from 10^{13} to 10^{14} microbial cells, and the microbial cell to human cell ratio of 1:1 in the colon, 3.8×10^{13} bacteria are estimated to exist (Kho and Lal 2018, Sender et al. 2016). The three major phyla recorded in GI tracts are *Firmicules*, *Bacterioidetes* and *Actinomycetes* (Tap et al. 2009), and their functional expansion has been found to be of 50–100 more genes that influence the host physiology (Hooper and Gordon 2001). The gut microbiome, for example, is vital for digestion and the generation of critical nutrients like vitamin K, whereas the skin microbiome protects the skin against dangerous microbes (Braat et al. 2006, Hooper and Gordon 2001, Kho and Lal 2018, Sender et al. 2016, Tap et al. 2009). Microbiome research has expanded in recent years due to an increased interest and the possible influence of these complex ecosystems on human health. The most important fields of microbiome study are shown in Figure 1.

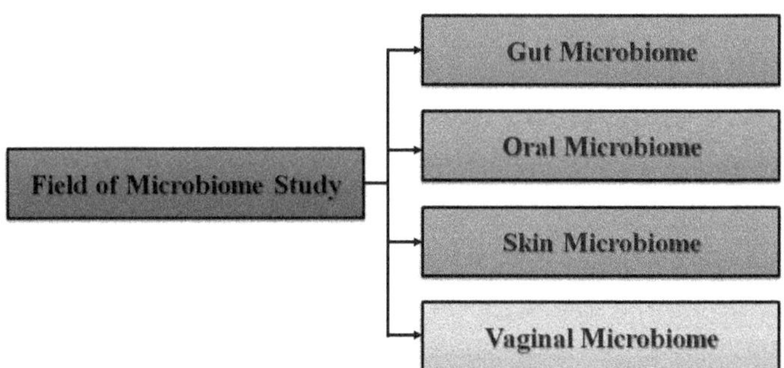

Figure 1. Important fields of microbiome study.

Gut Microbiome

The gut microbiome is one of the most well-studied microbiomes. Various studies have reported that the make-up of the gut microbiome can have a substantial influence on human health. Inadequate gut microbiome balance has been linked to a variety of health issues, including obesity, inflammatory bowel disease and colon cancer, etc. Recent studies have also discovered that the gut microbiota plays a role in immune system regulation as well as in the development of neurological illnesses, including autism and Parkinson's disease. The gut microbiota is associated with metabolic efficiency and host energy centres (Gill et al. 2006), and enzymatic degradation of unabsorbed complex carbohydrates and polymers (Salminen et al. 1998), thus contributing to 10% of daily dietary energy required for metabolism(Payne et al. 2012), and 70% of ATP production in the colon from microbially synthesised short chain fatty acids (SCFAs), respectively (Firmansyah et al. 1989, Donohoe et al. 2011). *Serratia marcescens, Eubacterium lentum, Bacteroides fragilis, Enterobacter agglomerans* and *Enterococcus faecium* are gut microbiota that produce vitamin K2 (menaquinone) anaerobically and are essential for lowering vascular calcification, hence lowering the risks of disorders like atherosclerosis and coronary heart diseases (cardiovascular disorders), and also elevating HDL and lowering cholesterol levels, etc. (Kawashima et al. 1997, Geleijnse et al. 2004).

Oral Microbiome

The microbes that reside inside the oral cavities are generally termed the oral microbiome – the second largest microbiome community in humans, after the gut microbiome. The soft and hard tissues of the oral cavity provide an environment for these microbiomes to colonise and form large communities. They form a bacterial biofilm at these places and provide an ideal environment for these micobiomes to flourish in the oral cavity. Various factors like temperature, pH and the surrounding environment make these microbiomes a suitable spot to colonise and multiply; in addition, the saliva present in the oral cavity keeps them hydrated and provides a supply of nutrients necessary for their multiplication (Dewhirst et al. 2010, Zhao et al. 2017). Significant research has been going on recently in the oral microbiome, with studies concentrating on the association between oral microbiome imbalances and illnesses such as periodontal disease, tooth decay and oral cancer (Hou et al. 2022).

Skin Microbiome

The skin microbiome plays a vital role in protecting the skin against dangerous microbes and preserving skin health. Skin problems such as acne, eczema and psoriasis have been linked to abnormalities in the skin microbiome, according to research. In general, the exterior surface of the human body is such that the skin acts as a protective barrier against the various types of environments and helps in the prevention of various foreign pathogenic particles entering, while simultaneously providing safety to the many microbes through commensalism (Byrd et al. 2018). The skin serves as the body's external connect with external factors and works as an invisible barrier to keep out foreign infections, while housing the commensal microbiota. The hardships that infections encounter when colonising human skin are additionally a result of the tough physical environment of the skin – dry, low in nutrients, and acidic. Nonetheless, the skin is home to a variety of microbes. The taxonomic diversity of microorganisms linked to the skin has been evaluated using amplicon and shotgun metagenomic DNA sequencing studies (Byrd et al. 2018, Grice and Segre 2011), from the group of organisms to the strain level. In addition, with an emphasis on *Propionibacterium acnes, Staphylococcus epidermidis* and *Staphylococcus aureus,* many researchers are uncovering more about skin microbial populations (particularly how they develop in health and disease), relationships among species, and how they affect the immune system.

Vaginal Microbiome

A recent study has focused on the significance of the vaginal microbiome in women's health, as well as the association between vaginal microbiome imbalances and illnesses such as bacterial vaginosis and urinary tract infections (Grice and Segre 2011).

Microbiome and Disease

Researchers are exploring the role of microbiomes in the genesis and progression of numerous illnesses, in addition to analysing individual microbiomes. An imbalance in the gut microbiota, for example, has been related to illnesses such as cardiovascular diseases and Type 2 diabetes, according to a study. Microbiome research is now a fast-expanding topic with considerable promise for improving human health (Grice and Segre 2011). In this chapter, we shall delve into their responsibilities and functions.

Microbiomes in Infection

The microbiome is a multifaceted population of bacteria found in different parts of the body, including the stomach, skin, mouth, lungs and other organs. In the setting of infection, the microbiome can act as a reservoir for potential infections or affect the immune response to disorders (Grice and Segre 2011). Microbiomes can act as disease reservoirs, influence immune responses, produce antimicrobial compounds, and compete for resources with infections. Microbiomes play an important role in infection by functioning as reservoirs for prospective pathogens (Grice and Segre 2011). There may be microbes, in the microbiome, that are generally innocuous or even beneficial, but can transform into opportunistic pathogens when the chance arises. For example, bacteria such as *Salmonella* species, *Clostridium difficile* and *Campylobacter* sps. can live in the gut microbiome and cause gastrointestinal infections when the stomach's natural barriers are overcome, such as during a compromised immune response or due to disturbances in the gut lining. Similarly, microorganisms such as *Streptococcus mutans* can be found in the oral microbiome and cause dental caries or tooth decay when the oral environment becomes unbalanced (Geleijnse et al. 2004).

Microbiomes can affect the immune response to diseases, in addition to acting as repositories for prospective pathogens. Beneficial bacteria in the microbiome can drive the generation of anti-inflammatory cytokines, boost immune cell growth, and improve the barrier function of the gut lining and other body surfaces. This boosts the immune system and protects the individual against illnesses. Certain bacteria strains in the gut microbiome, for example, have been demonstrated to induce the development of regulatory T cells, which are critical for immunological control and tolerance. However, dysbiosis or disturbance of the microbiome, which can occur as a result of antibiotic usage, poor nutrition, stress or other reasons, can result in an altered immune response, making the host more vulnerable to infections (Geleijnse et al. 2004, Kawashima et al. 1997).

Antimicrobial substances produced by microbiomes can limit the development of prospective infections. Certain kinds of bacteria in the microbiome, for example, create short-chain fatty acids (SCFAs), which have antibacterial qualities and aid in the maintenance of a healthy gut environment (Firmansyah et al. 1989, Donohoe et al. 2011). SCFAs reduce the pH of the stomach, making it unsuitable for the development of dangerous bacteria. Bacteriocins, which are produced by certain strains of bacteria in the oral microbiome and can inhibit the growth of harmful bacteria, and lactic acid, which is produced by certain bacteria in the vaginal microbiome and helps in maintaining an acidic environment that inhibits the growth of pathogenic microorganisms, are two other examples (Payne et al. 2012).

Furthermore, bacteria in the microbiome might fight for resources and space with prospective diseases, restricting their development and colonisation. This rivalry can take place via a variety of processes, including nutrition competition, the creation of inhibitory chemicals, and colonisation resistance. Beneficial bacteria in the gut microbiome, for example, might compete with pathogenic bacteria for nutrients such as carbohydrates, restricting the pathogens' access to resources (Bidell et al. 2022). They can also create inhibitory molecules that limit the growth of potential infections, such as bacteriocins or other antimicrobial substances. Furthermore, the healthy microbiome's colonisation resistance can prevent harmful bacteria from colonising the gut lining, by occupying niches and competing for adhesion sites (Bidell et al. 2022). Disruptions in the microbiome, on the other hand, might induce dysbiosis, which might reduce competition for resources and allow potential pathogens to thrive and cause diseases.

Broad-spectrum antibiotics, for example, can disturb the gut microbiome, causing a drop in beneficial bacteria and allowing opportunistic pathogens like *Clostridium difficile* to flourish and cause illnesses. Microbiomes are important in infection and they can serve as repositories for prospective infections, influence immune responses, create antimicrobial chemicals, and compete with diseases for resources. Maintaining a healthy and balanced microbiome is critical for infection prevention and general wellness (Bidell et al. 2022, Byrd et al. 2018, Grice and Segre 2011). Dysbiosis (imbalance) in humans impairs the normal functioning of the gut microbiota, leading to the dysregulated production of microbially derived products or metabolites that are harmful to the host; the associated infections and diseases are tabulated in Table 1. The other symptoms of gut dysbiosis are acid reflux, general malaise, fatigue, blood or mucus in the stool, constipation, intestinal pain, nausea or vomiting, etc. (Grice and Segre 2011, Bidell et al. 2022, Geleijnse et al. 2004, Hou et al. 2022).

Due to extrinsic factors, there is an imbalance due to the disproportionality of colonisation of gut microbiota, which thereby leads to an overgrowth of bacteria (more than 10^3 CFU/mL) in the upper gut with SIBO. Patients with SIBO have registered species of *Streptococcus, Acinetobacter, Kbelsiella, Enterococcus, pseudomonas, Staphylococcus,* etc., leading to altered GI motility (Kho and Lal 2018). SIBO creates several complications in the host, such as destruction of microvilli, enhanced epithelial inflammatory response, impaired absorption (Braat et al. 2006, Hooper and Gordon 2001, Tap et al. 2009, Sender et al. 2016), and other common symptoms such as abdominal discomfort, distension, bloating, deficiency of fat-soluble vitamins (D, E, A and K) (Gill et al. 2006), dysregulated fermentation of carbohydrates, and bacterial colonisation in the small intestine (with production of methane, hydrogen, carbon dioxide, etc.) (Firmansyah et al. 1989, Payne et al. 2012, Salminen et al. 1998).

Inflammatory Bowel Disease (IBD)

IBD is classified into two forms: *Crohn's* disease (CD) and ulcerative *colitis* (UC), which is a multifactorial, idiopathic, persistent disease that causes gastrointestinal (GI) inflammation (Lennard-Jones 1989). CD inflammation occurs in the restricted parts of the large intestine, and is estimated to occur in 1.4–2.2 million individuals in America and Europe.

Table 1. Gut microbiome and associated human diseases (Kho and Lal 2018).

Dysbiotic condition	**Infection**	**References**
Adverse increase in *E. coli*; depletion of *Clostridium leptum* and *Bifidobacterlum* sps.	Irritable Bowel Syndrome	(Duboc et al. 2012)
Increase in *Enterobacteriaceae* and *Bacteroides fragilis*	Inflammatory Bowel Disorders (IBD)	(Sokol et al. 2008, Png et al. 2010, Machiels et al. 2014)
Decrease in *Bifidobacterium* and increase in *Bacteroides – Prevotella* sps.	Celiac disease	(Tjellström et al. 2005, De Vadder et al. 2014)
High level of *Bacteroidetes* and decrease in *Actinobacteria* sps., *Furmicutes* and *Firmicetes* sps./*Bacteroidetes* sps. ratio	Type 1-diabetes	(Murri et al. 2013)
Increase in *Clostridium difficile*, and *C. difficile*/*Bifidobacteria* ratio	Atopic disease	(Kalliomäki et al. 2001)
Increase in *Blautia* sps. and *Proteobacteria* sps.	Systemic Lupus ErythemAtosus (SLF)	(Shi et al. 2014, Luo et al. 2018)
Decrease in *Odori bacter* sps. and *Alistipes* sps.	Rhematoid Arthritis (RA)	(Liu et al. 2013, Scher et al. 2013)
Increase in *Bacteroides fragilis*, *Fusobacterium* sps. and *Campylobacter* sps., and decrease in *Fascalibacterium* sps. and *Roseburia* sps.	Colorectal Cancer (CRC)	(Qin et al. 2012, Sato et al. 2014)
Increase in *Firmicutes* sps. and *Actinobacter* sps.	Obesity	(Wang et al. 2011, Wu et al. 2003, Turnbaugh et al. 2006)
Increase in *Clostridium* sps., *Bacteroidetes* sps., *Lactobacillus* sps., *Desulflovibrio* sps. and Decrease in *Bifidobacteria* sps.	Aulism Spectrum Disorder (ASD)	(Koliada et al. 2017, Song et al. 2008)
Increase in *Eggerthella* sps., *Gelria* sps., *Turicibacter* sps., *Anaerofilum* sps.	Depression	(Adams et al. 2011)
Increase in *Clostridium difficile* and decrease in *C. scindans*	Clostridium Difficile Infection (CDI)	(Kelly et al. 2016)
Increase in *Blautiacoprococcus* sps.	Parkinson's disease	(Theriot et al. 2014)
Increase in *Firmicetes* sps., *Proteobacteria* sps., *Actinobacteria* sps., and decrease in *Lactobacilli* sps.	Chronic Kidney Disease	(Keshavarzian et al. 2015, Vaziri et al. 2013)

The mechanism of pathogenesis between gut microbiota, the host and the environment is illustrated in Figure 2.

The aetiology of IBD is due to the hyper-responsiveness of T lymphocytes towards non-pathogenic antigens produced by the gut

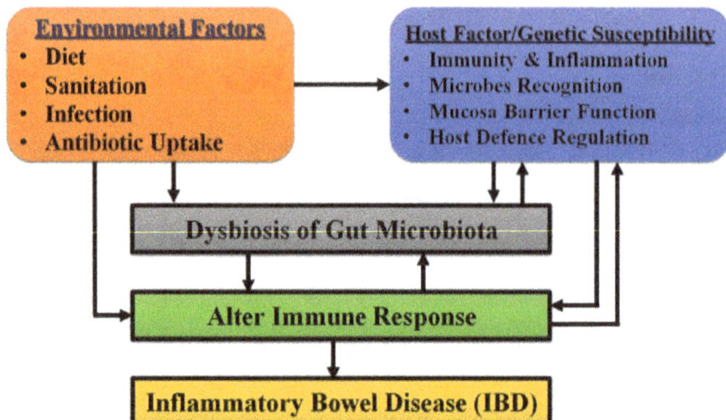

Figure 2. Mechanisms of pathogenesis between gut microbiota, host and environment.

microbiota (e.g., *Saccharomyces cerevisiae* and *Pseudomonas fluoresecens*) (Duchmann et al. 1999, Landers et al. 2002), which have distinct correlations with distinct clinical characteristics, disease onset and severity, etc., and affect gut barrier inflammation to different degrees (Vasiliauskas et al. 2000), leading to secondary gut inflammation. In IBD patients, observations have recorded very few gut *Firmicutes*, like *Faecalibacterium praunsnitzii* and *Roseburia* sps., with decreased anti-inflammatory cytokines, Short-Chain Fatty Acids (SCFA) deficiency in colonic barrier function, etc. (Sokol et al. 2008, Willing et al. 2010, Machiels et al. 2014). Colon biopsies have revealed a reduced mucus layer, leukocyte infiltration in the colon mucosa of IBD patients with UC (Willing et al. 2010), and elevation of mycolytic *Ruminococcus* sps. levels (Png et al. 2010). IBD patients have also recorded neurotoxic side effects on long-term follow-ups (Hyde et al. 1998, Genth et al. 2006).

Clostridium Difficile Infection (CDI)

C. difficile is a gram-positive bacterium belonging to Firmicutes commonly known as invaders of the gut (spore-producing anaerobes). In 2011, approximately 453,000 mortality cases were recorded in America (Lessa et al. 2015). Somesymptoms associated with CDI are sepsis, diarrhoea (due to the uptake of broad-spectrum antibiotics such as clindamycin, chloramphenicol, erythromycin, linzolid, etc.), and pseudomembranous colitis leading to death. The toxins produced by *C. difficile* are TcdA and TcdB (toxin A and toxin B), responsible for the cytoskeleton and the colonic epithelial barrier integrity (Genth et al. 2006, Bartlett 2002). Cholate derivatives (primary bile acid) serve as germinants for *C. difficile* spores, but deoxycholate (secondary bile acid) inhibits vegetative growth (Sorg and Sonenshein 2008). Increased growth of *C. difficile* exerts damage on the intestinal barrier, leading to severe inflammation and diarrhoea with intestinal ion absorption (Samuel et al. 2008, Theriot et al. 2014).

Celiac Disease

In Europe, 1% of the entire population is affected by celiac disease (a lifelong disease) (Mustalahti et al. 2010), with a 10.4 /1000-person mortality rate (Ludvigsson et al. 2009). Gut dysbiosis in celiac disease results in a decrease in the levels of *Atopobium, E. coli, Clostridium lituseburense, C. histolytica, Eubacterium rectal, C. coccoides,* etc. (Nadal et al. 2007, De Palma et al. 2010, Marasco et al. 2016). Celiac disease has also revealed a significant increase in total SCFs, i-butyrate, acetic acid, valetric acid, etc., indicating altered gut microbiota (Tjellström et al. 2005).

Autism Spectrum Disease (ASD)

Recently, according to a 2017 WHO report, autism spectrum disease has been classified under a multifarious group of neurobiological disorders and has a link between the gut microbiome and the brain, or neurodevelopmental disorders (such as mood disorders, depression, Alzheimer's, and PDs) (Christensen et al. 2015). Evidence of ASD was reported for 1 in 68 8-year-old children in America and 1 in 60 globally (Kho and Lal 2018); this disease is characterised by social, repetitive, communication deficit, and stereotyped behaviours (Constantino et al. 2000, Kim and Lord 2012). ASD is associated with tyrosine kinase MET signalling, which is important for brain development, immune response regulation, gastrointestinal health, etc. (Ieraci et al. 2002, Okunishi et al. 2005), and chronic abdominal pain (Carr and Owen-Deschryver 2007). Dysbiotic features in ASD patients show increased levels of *Clostridium, Bacteroides, Lactobacillus* and *Desulfovibrio* species and a decreased population of *Bifidobacteria* sps. (Song et al. 2008, Adams et al. 2011).

Obesity

Obesity is a global health hazard (WHO 2016), and it has been observed that the gut microbiome plays an important role in host metabolism regulation, elevated energy intake, excessive fat accumulation with raised body mass index (BMI > 30 kg/m^2), decreased energy expenditure, and low-grade inflammation (Donnelly et al. 2005, Cani et al. 2007). The colonic microbiome recorded in obese patients showed increased butyrate-producing *Firmicutes* and decreased *Bacteroidetes,* and the dysbiosis features were elevations of starch-degrading glycoside hydrolase and SCFs (Ley et al. 2006, Turnbaugh et al. 2006). Triacylglyceride accumulation and storage in adipocytes is also facilitated by the gut microbiota via lipoprotein lipase inhibitors (Bäckhed et al. 2004, Solinas et al. 2006). This is because Gram-negative gut bacteria cause metabolic endotoxemia (plasma LPS) and Bifidobacteria sps decrease, which may downregulate intestinal endotoxin (Griffiths et al. 2004).

Type2Diabetes

The microbiome has been reported to have pathophysiological effects on glucose metabolism (Gurung et al. 2020). T2D-negative bacterial microbiomes are species of *Akkermansia, Bacteroides, Roseburia, Bifidobacterium* and *Faecalibacterium,* and the positive genera are *Fusobacterium, Blautia, Ruminococcus,* etc., *Bacteroides intestinalis* and *Bacteroides vulgates* are decreased in T2D patients, and *Bacteroides stercoris* are higher in T2D patients after gastrectomy, which proves that Bacteroides play a beneficial role in glucose metabolism in humans (Hooper and Gordon 2001, Kawashima et al. 1997, Tjellström et al. 2005, Shi et al. 2014). Furthermore, the levels of *Faecalibacterium* species decrease due to different types of antidiabetic treatment (e.g., metformin) (Adams et al. 2011, Kelly et al. 2016, Theriot et al. 2014, Keshavarzian et al. 2015, Vaziri et al. 2013). Evidence of increased populations of *Lactobacillus salivaricus, L. acidophilus* and *L. gasseri* and decreased populations of *L. amylovorus* have also been reported in T2D patients. Other molecular mechanisms (metabolic diseases) associated with T2D patients are gut permeability, interaction with dietary constituents, glucose and lipid metabolism, overall energy homeostasis and inflammation (Gurung et al. 2020, Wu et al. 2010, Murphy et al. 2017, Zhang et al. 2013, Karlsson et al. 2013, Remely et al. 2014, Furet et al. 2010, Graessler et al. 2013, Yang et al. 2017). Pathobionts in T2D (*Fusobacterium nucleatum* and *Ruminococcus gnavis*) increase several inflammatory cytokines (Wu et al. 2010). Reduction in gut permeability and reduction in LPD production in T2D patients occur due to the expression of tight junction genes by *Bacteroides vulgaris* and *B. dorei* (gut microbiome) (Murphy et al. 2017, Zhang et al. 2013, Karlsson et al. 2013, Hall et al. 2017).

Other Dysbiosis

A 16S rRNA sequencing study of faecal microbiota has revealed the enrichment of *Bacteroides fragilus* (Wang et al. 2012, Wu et al. 2013), which in turn enhances the production of oncogenic *B. fragilis* enterotoxin (BFT), whose gene expression is significant in CRC patients – leading to colitis and colonic tumours. The type of arthritis (chronic inflammation) affecting the back, neck, and sometimes joints such as hips, knees, ankles or shoulders, is commonly known as ankylosing spondylitis (Liu et al. 2013, Qin et al. 2012). Postmenopausal women register an increased population of *Clostridium* and *Ruminococcus* species and a decreased amounts of Bacteroides (Vogt et al. 2017, Vemuri et al. 2019). They found that reproductively senescent female Sprague-Dawley rats showed a baseline elevation in the Firmicutes/Bacteroidates ratio, with a significant shift following stroke and decreased diversity. Interestingly, adult and middle-aged female rats showed increased populations of *Prevetella* and *Lactobacillus* species, with decreased SFCA and increased LPS levels (Park et al. 2020, Tiamani et al. 2022, Fulci et al. 2021), gut dysbiosis, biota-brain communications (Matijašić et al. 2020, Zárate et al. 2017) and decreased mucin production in female rats (Łoś and Węgrzyn 2012).

Virome in Infection

The virome is a collection of viruses found in and on the human body, including bacteriophages (viruses that infect bacteria), viruses that infect human cells, etc. The virome may also play a role in infections, both as a source of viral pathogens and as a regulator of the microbiome (Erez et al. 2017). Viromes, which are the totality of the genetic material of viruses found in different parts of the body, have become an important aspect of the study of infections. Viruses are the most populous and diverse group of microbes on earth, and they can have far-reaching consequences for the human host and other microbiomes. Viromes can play a role in both well-being and disease, from viral infections to the regulation of immune responses to interactions with other microbiomes.

The potential of viromes to cause viral infections in humans is one of their key functions in infections. Viruses can infect a variety of tissues and organs in the body, leading to a wide range of diseases such as respiratory infections, gastrointestinal infections, skin infections, etc. (Breitbart et al. 2008). Respiratory viruses, such as influenza and rhinoviruses, can cause common colds as well as more serious respiratory illnesses such as pneumonia. Gastrointestinal viruses, such as noroviruses and rotaviruses, can cause gastroenteritis, which is characterised by symptoms like vomiting and diarrhoea. Viruses can also cause serious health problems; for instance, hepatitis viruses cause liver infections, and HIV causes acquired immunodeficiency syndrome (AIDS) (Erez et al. 2017, Breitbart et al. 2008, Lim et al. 2015).

Viruses can not only cause viral infections but also modulate immune responses. For example, some viruses can directly infect immunological cells, causing activation or inhibition of the immune system. Viruses can also cause immunological responses by producing viral antigens that stimulate the immune system. This immunological response can be both beneficial and harmful to the host. On one hand, a strong immune response can help control and prevent viral infections, and, on the other hand, too active an immune response can cause inflammation, tissue damage and autoimmune reactions (Breitbart et al. 2008). In addition, viromes can affect the structure and function of other microbiomes in the body, such as the gut microbiome, the skin microbiome and the respiratory microbiome. Viruses in these microbiomes can infect and alter bacteria, archaea and other microbes, leading to changes in their abundance and diversity. For example, certain bacteriophages (viruses that infect bacteria) can alter the composition of the gut microbiome by selectively infecting and lysing certain bacteria, thereby affecting the balance of bacterial populations. This viral predation can have significant effects on the overall structure and function of the microbiome and, in turn, on the host's health (Erez et al. 2017, Breitbart et al. 2008).

Similarly, viromes can contribute to the development and spread of antibiotic resistance genes. Antibiotic resistance is a global public health problem, and the transmission of antibiotic resistance genes among bacteria is an important contributor to this problem (Erez et al. 2017, Lim et al. 2015, Huurre et al. 2008). Viruses can serve as vectors for the spread of antibiotic resistance genes among bacteria, a process known as horizontal gene transfer. This can lead to the spread of antibiotic resistance within the microbiome and between different body sites, contributing to the formation of multidrug-resistant bacterial strains and making it more difficult to treat diseases. Viromes can trigger viral infections, regulate immune responses, interact with other microbiomes, and promote the proliferation of antibiotic resistance genes. Understanding the intricate connections between viromes and the human host, as well as other microbiomes, is critical for expanding our understanding of infections and developing new techniques for the prevention, diagnosis and treatment of infectious diseases. Further study of viromes and their role in infections has the potential to transform our understanding of infectious diseases and lead to the development of innovative treatments (Breitbart et al. 2008).

Many viruses are present in the gut (e.g., *Mycoviridae* sps., *Podoviridae* sps., *Siphoviridae* sps.), the skin and the oral cavity (e.g., *Caudovirus* sps.), and they are acquired from birth or seeded from the maternal microbiome, later shaped by dietary habits (Reyes et al. 2010, Gregory et al. 2020, Manrique et al. 2016). About 3–2800 bacterial viruses are reported in 1 gram of faeces (Kim et al. 2011, Santiago-Rodriguez and Hollister 2019). The dynasmism of the human virome starts from delivery and varies with age, dietary intake, host's immunological status, environmental factors, drug intake, etc. (Santiago-Rodriguez and Hollister 2019). Large populations of DNA phages of the order Caudiovirales (e.g., *Mycoviridae, Siphovirdae, Podoviridae*) (Robinson and Pfeiffer 2014, Virgin 2014, Crapser et al. 2016) and eukaryotic viruses like *Anelloviridae T* and *Herpesviridae T* have been identified in infant stool samples (Roth et al. 2020); these populations are lower in old age. There are four types of interactions for phages within the host: (i) phagesin the lytic phase infect the cells, leading to host cell lysis; (ii) they inject their genomes into bacterial cells; (iii) the prophage stage leads to direct lytic growth; and (iv) pseudolysogenyloses interaction when the phage genome enters a bacterial cell (Ahnstedt and McCullough 2019, Virani et al. 2020). The human virome is a group of viruses such as bacteriophages or phages (Manwani et al. 2013), archaeal viruses (Yutin et al. 2021), and the various virus communities in the human gut (adult as well as infant),as illustrated in Table 2.

Double-stranded DNA viruses are responsible for infectious diseases (e.g., *Papillomaviridae* sps., *Herpesviridae* sps., *Adenoviridae* sps.); rotavirus (*Reoviridae* sps.), norovirus (*Calciviridae* sps.), and enterovirus (*Picornaviridae* sps.) are classified as pathogenic RNA viruses, noted for gastroenteritis in humans (Shkoporov et al. 2021). The human gut is

Table 2. Human virome in adults and infants (Reyes et al. 2010).

Type of virus	Genome	Adult	Infant
Eukaryotic virus	Single stranded DNA	*Anelloviridae* sps. *Parvoviridae* sps. *Circoviridae* sps.	*Geminiviridae* sps. *Nanoviridae* sps.
	Double stranded DNA	*Adenoviridae* sps. *Herpesviridae* sps. *Iridoviridae* sps. *Mimiviridae* sps.	*Adenoviridae* sps. *Polymaviridae* sps.
	Single stranded RNA	*Caliciviridae* sps. *Astroviridae* sps. *Retroviridae* sps. *Togaviridae* sps.	*Caliciviridae* sps. *Astroviridae* sps. *Tombusviridae* sps.
	Double stranded RNA	*Picobirnoviridae* sps. *Reoviridae* sps.	*Picobirnaviridae* sps. *Reoviridae* sps. *Chrysoviridae* sps.
Archaeal virus			*Lipothrixviridae* sps.
Bacteriophages	Single stranded DNA	*Inoviridae* sps. *Microviridae* sps.	*Inoviridae* sps. *Microviridae* sps. *Almaviridae* sps. *Novaviridae* sps.
	Double stranded DNA	*Myoviridae* sps. *Podoviridae* sps. *CrAssphages* sps. *Siphoviridae* sps.	*Siphoviridae* sps. *Corticoviridae* sps. *Picoviridae* sps. *Albertoviridae* sps.

invaded by a diverse group of viruses belonging to the order *Crassvirales* (CrAss-like phages) (Schwarza et al. 2010). Evidence of phase variation is found during the symbiosis between a lutic phage (CrAss-like phage CrAss001) and a bacterial host (Han et al. 2020). IBD patients register small circular viruses such as pepper mild motte virus (e.g., *Tobamovirus*) with 12.9% prevalence, and pepino mosaic virus (e.g., Potexvirus sps.) with 10.8% prevalence (Küry et al. 2018, Zuo et al. 2021).

Human endogenous reteroviruses (HERVs), which account for 8% of the human genome, are involved in the pathophysiological pathways of multiple sclerosis and amyotrophic lateral sclerosis (Zuo et al. 2021). SARS-COV-2 patients have revealed a varied dysbiosis in the gut; their faeces exhibit enriched bacteriophages and enterobacterphages, associated with gut inflammation and host interferon responses (Li et al. 2020), virulence-associated gene expression, etc. (Lopetuso et al. 2016, Liang et al. 2020). The cervical swab samples of HIV-infected women register the presence of Herpesviridae, Genomoviridae, Anelloviridae and Papillomaviridae (Hewson et al. 2010). Intestinal dysbiosis of IBD patients have shown phages of *Farcalibacterium prasnitizii* that played a role in the aetiology of the disease (Megremis et al. 2020), as well as infected Clostridiales, Alermonadales, etc., and an increased population of Retroviridae (Chen et al. 2014).

Respiratory Syncytical Virus (RSV) belonging to rhinoviruses presents a risk factor in asthma patients, due to mucin hypersecretion action (asthma exacerbation) (Damin et al. 2013). Asthma-affected children and pneumonia patients register RSV-B, RV-C, bocavirus and parovirus B-19, EJ-1, *Streptococcus* phage, RSV-B and RV-A (Emlet et al. 2020). Viruses account for 10–15% of the causative agents in all cancers, such as EBV, HPV, Merkel cell, polyomavirus, Kaposidsatcoma herpesvirus, T lymphotropic virus-1, Simian virus 40, etc. (Goel et al. 2006). HPV 18 is found in colorectal cancer patients from Asia and Europe,while HPV 16 is found in South American patients (Coelho et al. 2013, Dodi et al. 2021). T antigen (T-Ag), encoded by Human Polyomavirus 2 in colorectal cancer, is involved in oncogenesis, by inducing methylation of tumour suppressor gene promoters (Robinson et al. 2014, Freer et al. 2018); T-Ag also deregulates the Wnt signalling pathway through catenins. Anelloviridae are strongly associated with respiratory diseases as they influence both innate and adaptive immunity (Reyes et al. 2013) through pathogen-associated molecular pattern (PAMP) receptor interactions and activating the inflammasome. The above interaction resulted in lowering of T-lymphocyte and eosinophil counts (Canchaya et al. 2003). Caudoviralesphages aggravate colitis (immune response stimulation), alter specific gut bacterial species, and trigger inflammatory signalling (Canchaya et al. 2003).

Relationship Between Microbiomes, Viromes, and Human Health

Human health depends on microbiomes and viromes. They are important for regulating the immune system, preventing colonisation by pathogenic bacteria, and maintaining a healthy gut barrier. A good gut barrier is necessary because it prevents toxins and bacteria in the gut from entering the bloodstream, which can cause systemic inflammation and diseases such as autoimmune disorders. When the microbiome and the virome are altered, it can cause a variety of health problems. This disruption can increase the risk for infections, inflammatory bowel disease, allergies and other conditions. Maintaining a healthy balance of the microbiome and the virome is critical for human health and preventing infections (Kalliomäki et al. 2001, Scher et al. 2013, Erez et al. 2017, Zárate et al. 2017).

Manipulating Microbiomes and Viromes to Prevent or Treat Infections

Manipulating microbiomes and viromes, andthe genetic material of viruses, to prevent or treat infections, is an exciting field of study that holds enormous potential for revolutionising our approach towards infectious diseases. Probiotics are one method for modifying microbiomes and viromes for disease prevention or therapy (Hemarajata and Versalovic 2013). Probiotics are living bacteria that, when consumed in sufficient quantities, provide health advantages to the host. They can be used to modify the microbiome's make-up and function by increasing the development of beneficial microorganisms that can compete with harmful microbes, thus strengthening the gut barrier and modulating the immune response. Certain *Lactobacillus* and *Bifidobacterium* strains, for instance, have been shown to prevent or reduce the severity of gastrointestinal infections such as rotavirus and *Clostridium difficile* infections, by competing for resources with pathogenic microbes and modulating the immune response in the gut (Hemarajata and Versalovic 2013).

FMT, also known as faecal bacteriotherapy, involves transferring faeces from a healthy donor to a recipient with a disordered or dysbiotic microbiome. FMT has been demonstrated to be extremely successful in the treatment of recurrent *Clostridium difficile* infections, a dangerous and sometimes debilitating gastrointestinal condition that can be difficult to cure with antibiotics alone. FMT works by replenishing the recipient's microbiome, which can outcompete and restrict the growth of harmful microorganisms, and help restore gut health.

Viromes can also be altered to prevent or cure infections. Phage treatment is one strategy that includes employing viruses, known as bacteriophages, to target and eliminate specific bacteria that cause infections. Bacteriophages are very precise in their function, attacking just certain bacteria while ignoring good bacteria, and they can be used to selectively eradicate pathogenic bacteria from the microbiome (Hemarajata and Versalovic 2013, Zárate et al. 2017). Phage therapy has yielded encouraging results in the treatment of bacterial infections, particularly those caused by antibiotic-resistant bacteria, and is being intensively explored as a potential antibiotic replacement.

Another possibility, in addition to direct manipulation of microbiomes and viromes, is the use of prebiotics, which are substances that selectively encourage the development or activity of beneficial microorganisms in the microbiome. Prebiotics are indigestible fibres that can feed some beneficial bacteria, boosting their development and activity. It is possible to selectively stimulate the growth of beneficial microorganisms in the microbiome, by administering the proper prebiotics, which can help prevent the colonisation of harmful microbes and increase the immune response to infections (Hemarajata and Versalovic 2013).

On top of that, advances in gene editing technologies like CRISPR-Cas have opened new avenues for modifying microbiomes and viromes to prevent or treat ailments. CRISPR-Cas systems can modify the genetic material of certain bacteria or viruses, allowing for focused regulation of their activities (Strathdee et al. 2023). CRISPR-Cas, for example, may be used to engineer bacteria in the microbiome to make antimicrobial peptides capable of killing or inhibiting the development of dangerous bacteria. CRISPR-Cas may also be used to modify bacteriophages so that they target and destroy specific bacteria, giving a very precise and personalised method for treating infections caused by bacteria. Regardless of the tremendous promise of modifying microbiomes and viromes for disease prevention or treatment, hurdles and ethical concerns still remain to be solved. Because of the intricate and dynamic nature of microbiomes and viromes, as well as the possibility of unforeseen consequences, thorough study and monitoring are required to assure safety and efficacy (Strathdee et al. 2023). Antiviral medicines can be used to treat diseases caused by viromes, such as herpes simplex and hepatitis B. Antiviral medications operate by preventing virus replication, which can aid in infection control. Vaccines can also be used to prevent ailments including influenza, human papillomavirus, and COVID-19 (Strathdee et al. 2023).

Future Directions and Challenges

While the ongoing research on microbiomes and viromes has provided valuable insights, several challenges still need to be addressed. The vast diversity and complexity of microbial communities, along with inter-individual variations, present significant hurdles in deciphering their roles in health and disease. Standardisation of sampling techniques, data analysis, and ethical considerations are essential to ensure robust and reproducible research outcomes. Additionally, long-term safety assessments of microbiome- and virome-based interventions are necessary before widespread clinical applications.

Further research is needed to understand the intricate interactions between the different components of microbiomes and viromes. This includes investigating the influence of viruses on the behaviour and function of bacteria, as well as the crosstalk between viruses and the host immune system. Unravelling these complex relationships will provide valuable insights into disease development and potential therapeutic targets. Long-term, longitudinal studies are crucial to track changes in the microbiome and the virome over time, and their association with health and disease. This approach will help identify biomarkers for disease risk, track the progression of infections, and assess the efficacy of therapeutic interventions. Integrating data from various 'omics' technologies, such as genomics, metagenomics, transcriptomics and proteomics, will provide a more comprehensive understanding of the microbiome and the virome. Combining these approaches will allow researchers to identify not only microbial compositions but also functional pathways and interactions, providing a more holistic view of their impact on human health. Developing targeted interventions to modulate microbiomes and viromes holds great promise. This includes strategies like faecal microbiota transplantation (FMT), where the transfer of healthy microbiota from a donor to a recipient can restore the microbial balance and treat certain diseases. Additionally, the development of phage-based therapies, prebiotics, probiotics and postbiotics offers exciting avenues for the manipulation of microbiomes and viromes, to prevent and treat infections.

Ensuring standardised protocols and methodologies for sample collection, storage and analysis is crucial for robust and reproducible research outcomes. This is particularly challenging due to the inherent variability of microbiomes and viromes among individuals and populations. Establishing standardised guidelines will enhance the comparability of studies and facilitate the translation of research findings into clinical applications. Analysing vast amounts of complex microbiome and virome data requires advanced bioinformatics tools and computational approaches. Developing robust algorithms for data analysis, integration and interpretation is essential for extracting meaningful insights from large-scale datasets and identifying relevant microbial signatures or patterns associated with diseases. As research progresses, ethical considerations become increasingly important. Safeguarding participant privacy, informed consent, and ensuring equitable access to emerging therapies are critical for maintaining public trust and ensuring that ethical standards are met in microbiome- and virome-based research. Before widespread clinical applications of microbiome- and virome-based interventions, rigorous evaluation of long-term safety profiles, and regulatory approval are necessary. Understanding the potential risks, such as unintended consequences or adverse effects, is crucial to ensure patient safety and the successful translation of research findings into clinical practice.

Conclusion

Microbiomes and viromes, which are made up of trillions of microorganisms such as bacteria, viruses, fungus and other microbes, are important in health and illness, including infections. Manipulation of microbiomes and viromes to prevent or treat infections is an active field of study that holds considerable potential for revolutionising our approach towards infectious diseases. The future of research in microbiome- and virome-based infections holds immense potential for advancements in disease prevention, diagnosis and treatment. To change the composition and the functions of microbiomes and viromes for infection prevention or treatment, approaches such as probiotics, faecal microbiota transplantation (FMT), phage therapy, prebiotics, and gene editing technologies such as CRISPR-Cas can be utilised. To assure the safety and efficacy of modifying microbiomes and viromes for infection management, however, rigorous research, monitoring, and ethical considerations are crucial. Hence, although microbiomes and viromes are essential for human health, their disruption can lead to infections and other health problems. Maintaining a healthy balance of microbiomes and viromes is crucial to promoting human health and preventing infections. By addressing the challenges of standardisation, data analysis, ethics, and safety, researchers can overcome barriers and unlock the full potential of these microbial ecosystems. Continued exploration of microbiomes and viromes will lead to transformative discoveries that have the potential to revolutionisehealthcare and improve patient outcomes in the years to come.

References

Adams, J. B., Johansen, L. J., Powell, L. D., Quig, D. and Rubin, R. A. (2011). Gastrointestinal flora and gastrointestinal status in children with autism—comparisons to typical children and correlation with autism severity. BMC Gastroenterol., 11: 22.

Ahnstedt, H. and McCullough, L. D. (2019). The impact of sex and age on T cell immunity and ischemic stroke outcomes. Cell Immunol., 345: 103960.

Bäckhed, F., Ding, H., Wang, T., Hooper, L. V., Koh, G. Y., Nagy, A. and Gordon, J. I. (2004). The gut microbiota as an environmental factor that regulates fat storage. Proc. Natl. Acad. Sci. U S A, 101(44), 15718–15723.

Bartlett, J. G. (2002). Clinical practice. Antibiotic-associated diarrhea. N. Engl. J. Med., 346(5): 334–9.

Bidell, M. R., Hobbs, A. L. V. and Lodise, T. P. (2022). Gut microbiome health and dysbiosis: A clinical primer. Pharmacotherapy, 42(11): 849–857.

Braat, H., Rottiers, P., Hommes, D. W., Huyghebaert, N., Remaut, E., Remon, J. P. and Steidler, L. (2006). A phase I trial with transgenic bacteria expressing interleukin-10 in Crohn's disease. Clin. Gastroenterol. Hepatol., 4(6): 754–9.

Breitbart, M., Haynes, M., Kelley, S., Angly, F., Edwards, R. A., Felts, B. and Rohwer, F. (2008). Viral diversity and dynamics in an infant gut. Res. Microbiol., 159(5): 367–73.

Byrd, A. L., Belkaid, Y. and Segre, J. A. (2018). The human skin microbiome. Nat. Rev. Microbiol., 16(3): 143–155.

Canchaya, C., Fournous, G., Chibani-Chennoufi, S., Dillmann, M. L. and Brüssow, H. (2003). Phage as agents of lateral gene transfer. Curr. Opin. Microbiol., 6(4): 417–24.

Cani, P. D., Amar, J., Iglesias, M. A., Poggi, M., Knauf, C., Bastelica, D. and Burcelin, R. (2007). Metabolic endotoxemia initiates obesity and insulin resistance. Diabetes, 56(7): 1761–72.

Carr, E. G. and Owen-Deschryver, J. S. (2007). Physical illness, pain, and problem behavior in minimally verbal people with developmental disabilities. J. Autism. Dev. Disord., 37(3): 413–24.

Chen, Y., Williams, V., Filippova, M., Filippov, V. and Duerksen-Hughes, P. (2014). Viral carcinogenesis: Factors inducing DNA damage and virus integration. Cancers (Basel), 6(4): 2155–86.

Christensen, K. R., Steenholdt, C., Buhl, S. S., Ainsworth, M. A., Thomsen, O. Ø. and Brynskov, J. (2015). Systematic information to health-care professionals about vaccination guidelines improves adherence in patients with inflammatory bowel disease in anti-TNFα therapy. Am. J. Gastroenterol., 110(11): 1526–32.

Coelho, T. R., Gaspar, R., Figueiredo, P., Mendonça, C., Lazo, P. A. and Almeida, L. (2013). Human JC polyomavirus in normal colorectal mucosa, hyperplastic polyps, sporadic adenomas, and adenocarcinomas in Portugal. J. Med. Virol., 85(12): 2119–27.

Constantino, J. N., Przybeck, T., Friesen, D. and Todd, R. D. (2000). Reciprocal social behavior in children with and without pervasive developmental disorders. J. Dev. Behav. Pediatr., 21(1): 2–11.

Crapser, J., Ritzel, R., Verma, R., Venna, V. R., Liu, F., Chauhan, A. and McCullough, L. D. (2016). Ischemic stroke induces gut permeability and enhances bacterial translocation leading to sepsis in aged mice. Aging (Albany NY), 8(5): 1049–63.

Damin, D. C., Ziegelmann, P. K. and Damin, A. P. (2013). Human papillomavirus infection and colorectal cancer risk: A meta-analysis. Colorectal. Dis., 15(8): e420-8.

De Palma, G., Nadal, I., Medina, M., Donat, E., Ribes-Koninckx, C., Calabuig, M. and Sanz, Y. (2010). Intestinal dysbiosis and reduced immunoglobulin-coated bacteria associated with coeliac disease in children. BMC Microbiol., 10(1): 1–7.

De Vadder, F., Kovatcheva-Datchary, P., Goncalves, D., Vinera, J., Zitoun, C., Duchampt, A. and Mithieux, G. (2014). Microbiota-generated metabolites promote metabolic benefits via gut-brain neural circuits. Cell, 156(1): 84–96.

Dewhirst, F. E., Chen, T., Izard, J., Paster, B. J., Tanner, A. C., Yu, W. H. and Wade, W. G. (2010). The human oral microbiome. Journal of bacteriology, 192(19): 5002–5017.

Dodi, G., Attanasi, M., Di Filippo, P., Di Pillo, S. and Chiarelli, F. (2021). Virome in the lungs: The role of anelloviruses in childhood respiratory diseases. Microorganisms, 9(7).

Donnelly, K. L., Smith, C. I., Schwarzenberg, S. J., Jessurun, J., Boldt, M. D. and Parks, E. J. (2005). Sources of fatty acids stored in liver and secreted via lipoproteins in patients with nonalcoholic fatty liver disease. J. Clin. Invest., 115(5): 1343–51.

Donohoe, D. R., Garge, N., Zhang, X., Sun, W., O'Connell, T. M., Bunger, M. K. and Bultman, S. J. (2011). The microbiome and butyrate regulate energy metabolism and autophagy in the mammalian colon. Cell. Metab., 13(5): 517–26.

Duboc, H., Rainteau, D., Rajca, S., Humbert, L., Farabos, D., Maubert, M. and Sabaté, J. M. (2012). Increase in fecal primary bile acids and dysbiosis in patients with diarrhea-predominant irritable bowel syndrome. Neurogastroenterol. Motil., 24(6): 513–20, e246-7.

Duchmann, R., May, E., Heike, M., Knolle, P., Neurath, M. and Meyer zum Büschenfelde, K. H. (1999). T cell specificity and cross reactivity towards enterobacteria, bacteroides, bifidobacterium, and antigens from resident intestinal flora in humans. Gut, 44(6): 812–8.

Emlet, C., Ruffin, M. and Lamendella, R. (2020). Enteric virome and carcinogenesis in the gut. Dig. Dis. Sci., 65(3): 852–864.

Erez, Z., Steinberger-Levy, I., Shamir, M., Doron, S., Stokar-Avihail, A., Peleg, Y. and Sorek, R. (2017). Communication between viruses guides lysis–lysogeny decisions. Nature, 541(7638): 488–493.

Firmansyah, A., Penn, D. and Lebenthal, E. (1989). Isolated colonocyte metabolism of glucose, glutamine, n-butyrate, and beta-hydroxybutyrate in malnutrition. Gastroenterology, 97(3): 622–9.

Freer, G., Maggi, F., Pifferi, M., Di Cicco, M. E., Peroni, D. G. and Pistello, M. (2018). The virome and its major component, anellovirus, a convoluted system molding human immune defenses and possibly affecting the development of asthma and respiratory diseases in childhood. Front. Microbiol., 9: 686.

Fulci, V., Stronati, L., Cucchiara, S., Laudadio, I. and Carissimi, C. (2021). Emerging roles of gut virome in pediatric diseases. Int. J. Mol. Sci., 22(8).

Furet, J. P., Kong, L. C., Tap, J., Poitou, C., Basdevant, A., Bouillot, J. L. and Clément, K. (2010). Differential adaptation of human gut microbiota to bariatric surgery–induced weight loss: links with metabolic and low-grade inflammation markers. Diabetes, 59(12): 3049–3057.

Geleijnse, J. M., Vermeer, C., Grobbee, D. E., Schurgers, L. J., Knapen, M. H., Van Der Meer, I. M. and Witteman, J. C. (2004). Dietary intake of menaquinone is associated with a reduced risk of coronary heart disease: the Rotterdam Study. The Journal of nutrition, 134(11): 3100–3105.

Genth, H., Huelsenbeck, J., Hartmann, B., Hofmann, F., Just, I. and Gerhard, R. (2006). Cellular stability of Rho-GTPases glucosylated by Clostridium difficile toxin B. FEBS Lett., 580(14): 3565–9.

Gill, S. R., Pop, M., DeBoy, R. T., Eckburg, P. B., Turnbaugh, P. J., Samuel, B. S. and Nelson, K. E. (2006). Metagenomic analysis of the human distal gut microbiome. Science, 312(5778): 1355–1359.

Goel, A., Li, M. S., Nagasaka, T., Shin, S. K., Fuerst, F., Ricciardiello, L. and Boland, C. R. (2006). Association of JC virus T-antigen expression with the methylator phenotype in sporadic colorectal cancers. Gastroenterology, 130(7): 1950–1961.

Graessler, J., Qin, Y., Zhong, H., Zhang, J., Licinio, J., Wong, M. L. and Bornstein, S. R. (2013). Metagenomic sequencing of the human gut microbiome before and after bariatric surgery in obese patients with type 2 diabetes: correlation with inflammatory and metabolic parameters. Pharmacogenomics J., 13(6): 514–22.

Gregory, A. C., Zablocki, O., Zayed, A. A. Howell, A., Bolduc, B. and Sullivan, M. B. (2020). The gut virome database reveals age-dependent patterns of virome diversity in the human gut. Cell Host Microbe., 28(5): 724–740.e8.

Grice, E. A. and Segre, J. A. (2011). The skin microbiome. Nat. Rev. Microbiol., 9(4): 244–53.

Griffiths, E. A., Duffy, L. C., Schanbacher, F. L., Qiao, H., Dryja, D., Leavens, A. and Ogra, P. L. (2004). *In vivo* effects of bifidobacteria and lactoferrin on gut endotoxin concentration and mucosal immunity in Balb/c mice. Dig. Dis. Sci., 49(4): 579–89.

Gurung, M., Li, Z., You, H., Rodrigues, R., Jump, D. B., Morgun, A. and Shulzhenko, N. (2020). Role of gut microbiota in type 2 diabetes pathophysiology. EBioMedicine, 51.102590.

Hall, A. B., Yassour, M., Sauk, J., Garner, A., Jiang, X., Arthur, T. and Huttenhower, C. (2017). A novel Ruminococcus gnavus clade enriched in inflammatory bowel disease patients. Genome Med., 9(1): 103.

Han Yang, Zhilong Jia, Jinlong Shi, Weidong Wang and Kunlun He. (2020). The active lung microbiota landscape of COVID-19 patients (preprint).

Hemarajata, P. and Versalovic, J. (2013). Effects of probiotics on gut microbiota: Mechanisms of intestinal immunomodulation and neuromodulation. Therap. Adv. Gastroenterol., 6(1): 39–51.

Hewson, C. A., Haas, J. J., Bartlett, N. W., Laza-Stanca, V., Kebadze, T., Caramori, G. and Johnston, S. L. (2010). Rhinovirus induces MUC5AC in a human infection model and *in vitro* via NF-κB and EGFR pathways. Eur. Respir. J., 36(6): 1425–35.

Hooper, L. V. and Gordon, J. I. (2001). Commensal host-bacterial relationships in the gut. Science, 292(5519): 1115–8.

Hou, K., Wu, Z. X., Chen, X. Y., Wang, J. Q., Zhang, D., Xiao, C., Zhu, D., Koya, J. B., Wei, L., Li, J. and Chen, Z. S. (2022). Microbiota in health and diseases. Signal Transduction and Targeted Therapy, 7(1): 135. https://doi.org/10.1038/s41392-022-00974-4.

Huurre, A., Kalliomäki, M., Rautava, S., Rinne, M., Salminen, S. and Isolauri, E. (2008). Mode of delivery-effects on gut microbiota and humoral immunity. Neonatology, 93(4): 236–40.

Hyde, G. M., Thillainayagam, A. V. and Jewell, D. P. (1998). Intravenous cyclosporin as rescue therapy in severe ulcerative colitis: time for a reappraisal? Eur. J. Gastroenterol. Hepatol., 10(5): 411–3.

Ieraci, A., Forni, P. E. and Ponzetto, C. (2002). Viable hypomorphic signaling mutant of the Met receptor reveals a role for hepatocyte growth factor in postnatal cerebellar development. Proc. Natl. Acad. Sci. U S A, 99(23): 15200–5.

Kalliomäki, M., Kirjavainen, P., Eerola, E., Kero, P., Salminen, S. and Isolauri, E. (2001). Distinct patterns of neonatal gut microflora in infants in whom atopy was and was not developing. J. Allergy Clin. Immunol., 107(1): 129–34.

Karlsson, F. H., Tremaroli, V., Nookaew, I., Bergström, G., Behre, C. J., Fagerberg, B. and Bäckhed, F. (2013). Gut metagenome in European women with normal, impaired and diabetic glucose control. Nature, 498(7452): 99–103.

Hidetoshi, K., Yoshikage, N., Yoshio, M., Junichi, N., Taneo, F., Saburo, M. and Tetsuya, N. (1997). Effects of vitamin K2 (menatetrenone) on atherosclerosis and blood coagulation in hypercholesterolemic rabbits. Jpn. J. Pharmacol., 75(2): 135–43.

Kelly, J. R., Borre, Y., O'Brien, C., Patterson, E., El Aidy, S., Deane, J. and Dinan, T. G. (2016). Transferring the blues: depression-associated gut microbiota induces neurobehavioural changes in the rat. J. Psychiatr. Res., 82: 109–18.

Keshavarzian, A., Green, S. J., Engen, P. A., Voigt, R. M., Naqib, A., Forsyth, C. B. and Shannon, K. M. (2015). Colonic bacterial composition in Parkinson's disease. Movement Disorders, 30(10), 1351–1360.

Kho Zhi Y. and Sunil K. Lal. (2018). The Human Gut Microbiome—A Potential Controller of Wellness and Disease, 9.

Kim, M. S., Park, E. J., Roh, S. W. and Bae, J. W. (2011). Diversity and abundance of single-stranded DNA viruses in human feces. Appl. Environ. Microbiol., 77(22): 8062–70.

Kim, S. H. and Lord, C. (2012). New autism diagnostic interview-revised algorithms for toddlers and young preschoolers from 12 to 47 months of age. J. Autism. Dev. Disord., 42(1): 82–93.

Koliada, A., Syzenko, G., Moseiko, V., Budovska, L., Puchkov, K., Perederiy, V. and Vaiserman, A. (2017). Association between body mass index and Firmicutes/Bacteroidetes ratio in an adult Ukrainian population. BMC Microbiol., 17(1): 1–6.

Küry, P., Nath, A., Créange, A., Dolei, A., Marche, P., Gold, J. and Perron, H. (2018). Human endogenous retroviruses in neurological diseases. Trends Mol. Med., 24(4): 379–394.

Landers, C. J., Cohavy, O., Misra, R., Yang, H., Lin, Y. C., Braun, J. and Targan, S. R. (2002). Selected loss of tolerance evidenced by Crohn's disease–associated immune responses to auto-and microbial antigens. Gastroenterology, 123(3): 689–699.

Lennard-Jones, J. E. (1989). Classification of inflammatory bowel disease. Scand. J. Gastroenterol. Suppl 170: 2–6; discussion 16-9.

Lessa, F. C., Mu, Y., Bamberg, W. M., Beldavs, Z. G., Dumyati, G. K., Dunn, J. R. and McDonald, L. C. (2015). Burden of Clostridium difficile infection in the United States. N. Engl. J. Med., 372(9): 825–34.

Ley, R. E., Turnbaugh, P. J., Klein, S. and Gordon, J. I. (2006). Microbial ecology: Human gut microbes associated with obesity. Nature, 444(7122): 1022–3.

Li, Y., Altan, E., Pilcher, C., Hartogensis, W., Hecht, F. M., Deng, X. and Delwart, E. (2020). Semen virome of men with HIV on or off antiretroviral treatment. Aids, 34(6): 827–832.

Liang, G., Conrad, M. A., Kelsen, J. R., Kessler, L. R., Breton, J., Albenberg, L. G. and Bushman, F. D. (2020). Dynamics of the stool virome in very early-onset inflammatory bowel disease. J. Crohns Colitis, 14(11): 1600–1610.

Lim, E. S., Zhou, Y., Zhao, G., Bauer, I. K., Droit, L., Ndao, I. M. and Holtz, L. R. (2015). Early life dynamics of the human gut virome and bacterial microbiome in infants. Nat. Med. 21(10): 1228–34.

Liu, X., Zou, Q., Zeng, B., Fang, Y. and Wei, H. (2013). Analysis of fecal Lactobacillus community structure in patients with early rheumatoid arthritis. Curr. Microbiol., 67(2): 170–6.

Lopetuso, L. R., Ianiro, G., Scaldaferri, F., Cammarota, G. and Gasbarrini, A. (2016). Gut virome and inflammatory bowel disease. Inflamm. Bowel Dis., 22(7): 1708–12.

Łoś, M. and Węgrzyn, G. (2012). Pseudolysogeny. Adv. Virus Res., 82: 339–49.

Ludvigsson, J. F., Montgomery, S. M., Ekbom, A., Brandt, L. and Granath, F. (2009). Small-intestinal histopathology and mortality risk in celiac disease. Jama, 302(11): 1171–8.

Luo, X. M., Edwards, M. R., Mu, Q., Yu, Y., Vieson, M. D., Reilly, C. M., Ahmed, S. A. and Bankole, A. A. (2018). Gut microbiota in human systemic lupus erythematosus and a mouse model of lupus. Applied and Environmental Microbiology, 84(4): e02288-17. https://doi.org/10.1128/AEM.02288-17.

Machiels, K., Joossens, M., Sabino, J., De Preter, V., Arijs, I., Eeckhaut, V. and Vermeire, S. (2014). A decrease of the butyrate-producing species Roseburia hominis and Faecalibacterium prausnitzii defines dysbiosis in patients with ulcerative colitis. Gut, 63(8), 1275–1283.

Manrique, P., Bolduc, B., Walk, S. T., van der Oost, J., de Vos, W. M. and Young, M. J. (2016). Healthy human gut phageome. Proc. Natl. Acad. Sci. U S A, 113(37): 10400–5.

Manwani, B., Liu, F., Scranton, V., Hammond, M. D., Sansing, L. H. and McCullough, L. D. (2013). Differential effects of aging and sex on stroke induced inflammation across the lifespan. Exp. Neurol., 249: 120–31.

Marasco, G., Di Biase, A. R., Schiumerini, R., Eusebi, L. H., Iughetti, L., Ravaioli, F. and Festi, D. (2016). Gut microbiota and celiac disease. Dig. Dis. Sci., 61(6): 1461–72.

Matijašić, M., Meštrović, T., Paljetak, H. Č., Perić, M., Barešić, A. and Verbanac, D. (2020). Gut microbiota beyond bacteria-mycobiome, virome, archaeome, and eukaryotic parasites in IBD. Int. J. Mol. Sci., 21(8).

Megremis, S., Constantinides, B., Xepapadaki, P., Bachert, C., Neurath-Finotto, S., Jartti, T. and Papadopoulos, N. (2020). Bacteriophage deficiency characterizes respiratory virome dysbiosis in childhood asthma. BioRxiv, 2020-08. 04.236067.

Murphy, R., Tsai, P., Jüllig, M., Liu, A., Plank, L. and Booth, M. (2017). Differential changes in gut microbiota after gastric bypass and sleeve gastrectomy bariatric surgery vary according to diabetes remission. Obes. Surg., 27(4): 917–925.

Murri, M., Leiva, I., Gomez-Zumaquero, J. M., Tinahones, F. J., Cardona, F., Soriguer, F. and Queipo-Ortuño, M. I. (2013). Gut microbiota in children with type 1 diabetes differs from that in healthy children: a case-control study. BMC Med., 11: 46.

Mustalahti, K., Catassi, C., Reunanen, A., Fabiani, E., Heier, M., McMillan, S. and members of the Coeliac EU Cluster, Epidemiology. (2010). The prevalence of celiac disease in Europe: results of a centralized, international mass screening project. Ann. Med., 42(8): 587–95.

Nadal, I., Donant, E., Ribes-Koninckx, C., Calabuig, M. and Sanz, Y. (2007). Imbalance in the composition of the duodenal microbiota of children with coeliac disease. J. Med. Microbiol., 56(Pt 12): 1669–1674.

Okunishi, K., Dohi, M., Nakagome, K., Tanaka, R., Mizuno, S., Matsumoto, K. and Yamamoto, K. (2005). A novel role of hepatocyte growth factor as an immune regulator through suppressing dendritic cell function. J. Immunol., 175(7): 4745–53.

Park, M. J., Pilla, R., Panta, A., Pandey, S., Sarawichitr, B., Suchodolski, J. and Sohrabji, F. (2020). Reproductive senescence and ischemic stroke remodel the gut microbiome and modulate the effects of estrogen treatment in female rats. Transl. Stroke Res., 11(4): 812–830.

Payne, A. N., Chassard, C., Banz, Y. and Lacroix, C. (2012). The composition and metabolic activity of child gut microbiota demonstrate differential adaptation to varied nutrient loads in an *in vitro* model of colonic fermentation. FEMS Microbiol. Ecol., 80(3): 608–23.

Png, C. W., Lindén, S. K., Gilshenan, K. S., Zoetendal, E. G., McSweeney, C. S., Sly, L. I. and Florin, T. H. (2010). Mucolytic bacteria with increased prevalence in IBD mucosa augmentin vitroutilization of mucin by other bacteria. Am. J. Gastroenterol., 105(11): 2420–8.

Qin, J., Li, Y., Cai, Z., Li, S., Zhu, J., Zhang, F. and Wang, J. (2012). A metagenome-wide association study of gut microbiota in type 2 diabetes. Nature, 490(7418): 55–60.

Remely, M., Aumueller, E., Merold, C., Dworzak, S., Hippe, B., Zanner, J. and Haslberger, A. G. (2014). Effects of short chain fatty acid producing bacteria on epigenetic regulation of FFAR3 in type 2 diabetes and obesity. Gene, 537(1): 85–92.

Reyes, A., Haynes, M., Hanson, N., Angly, F. E., Heath, A. C., Rohwer, F. and Gordon, J. I. (2010). Viruses in the faecal microbiota of monozygotic twins and their mothers. Nature, 466(7304): 334–338.

Reyes, A., Wu, M., McNulty, N. P., Rohwer, F. L. and Gordon, J. I. (2013). Gnotobiotic mouse model of phage-bacterial host dynamics in the human gut. Proc. Natl. Acad. Sci. U S A, 110(50): 20236–41.

Robinson, C. M., Jesudhasan, P. R. and Pfeiffer, J. K. (2014). Bacterial lipopolysaccharide binding enhances virion stability and promotes environmental fitness of an enteric virus. Cell Host Microbe, 15(1): 36–46.

Robinson, C. M. and Pfeiffer, J. K. (2014). Viruses and the microbiota. Annu. Rev. Virol., 1: 55–69.

Roth, W. H., Cai, A., Zhang, C., Chen, M. L., Merkler, A. E. and Kamel, H. (2020). Gastrointestinal disorders and risk of first-ever ischemic stroke. Stroke, 51(12): 3577–3583.

Salminen, S., Bouley, C., Boutron, M. C., Cummings, J. H., Franck, A., Gibson, G. R. and Rowland, I. (1998). Functional food science and gastrointestinal physiology and function. Br. J. Nutr., 80 Suppl 1: S147–71.

Samuel, B. S., Shaito, A., Motoike, T., Rey, F. E., Backhed, F., Manchester, J. K. and Gordon, J. I. (2008). Effects of the gut microbiota on host adiposity are modulated by the short-chain fatty-acid binding G protein-coupled receptor, Gpr41. Proc. Natl. Acad. Sci. U S A, 105(43): 16767–72.

Santiago-Rodriguez, T. M. and Hollister, E. B. (2019). Human virome and disease: High-throughput sequencing for virus discovery, identification of phage-bacteria dysbiosis and development of therapeutic approaches with emphasis on the human gut. Viruses, 11(7).

Santiago-Rodriguez, T. M. and Hollister, E. B. (2020). Potential applications of human viral metagenomics and reference materials: considerations for current and future viruses. Appl. Environ. Microbiol., 86(22).

Sato, J., Kanazawa, A., Ikeda, F., Yoshihara, T., Goto, H., Abe, H. and Watada, H. (2014). Gut dysbiosis and detection of "live gut bacteria" in blood of Japanese patients with type 2 diabetes. Diabetes care, 37(8): 2343–2350.

Scher, J. U., Sczesnak, A., Longman, R. S., Segata, N., Ubeda, C., Bielski, C. and Littman, D. R. (2013). Expansion of intestinal Prevotella copri correlates with enhanced susceptibility to arthritis. Elife, 2: e01202.

Schwarza, D., Beuchbc, U., Bandteb, M., Fakhroab, A., Büttnerb, C. and Obermeierbd, C. (2010). Spread and interaction of Pepino mosaic virus (PepMV) and Pythium aphanidermatum in a closed nutrient solution recirculation system: Effects on tomato growth and yield.

Sender, R., Fuchs, S. and Milo, R. (2016). Revised estimates for the number of human and bacteria cells in the body. PLoS Biol., 14(8): e1002533.

Shi, L., Zhang, Z., Yu, A. M., Wang, W., Wei, Z., Akhter, E. and Sullivan, K. E. (2014). The SLE transcriptome exhibits evidence of chronic endotoxin exposure and has widespread dysregulation of non-coding and coding RNAs. PLoS One, 9(5): e93846.

Shkoporov, A. N., Khokhlova, E. V., Stephens, N., Hueston, C., Seymour, S., Hryckowian, A. J. and Hill, C. (2021). Long-term persistence of crAss-like phage crAss001 is associated with phase variation in Bacteroides intestinalis. BMC Biol., 19(1): 163.

Sokol, H., Pigneur, B., Watterlot, L., Lakhdari, O., Bermúdez-Humarán, L. G., Gratadoux, J. J. and Langella, P. (2008). Faecalibacterium prausnitzii is an anti-inflammatory commensal bacterium identified by gut microbiota analysis of Crohn disease patients. Proc. Natl. Acad. Sci. U S A, 105(43): 16731–6.

Solinas, G., Summermatter, S., Mainieri, D., Gubler, M., Montani, J. P., Seydoux, J. and Dulloo, A. G. (2006). Corticotropin-releasing hormone directly stimulates thermogenesis in skeletal muscle possibly through substrate cycling between *de novo* lipogenesis and lipid oxidation. Endocrinology, 147(1): 31–38.

Song, H. J., Shim, K. N., Jung, S. A., Choi, H. J., Lee, M. A., Ryu, K. H. and Yoo, K. (2008). Antibiotic-associated diarrhea: candidate organisms other than Clostridium difficile. Korean J. Intern. Med., 23(1): 9–15.

Sorg, J. A. and Sonenshein, A. L. (2008). Bile salts and glycine as cogerminants for Clostridium difficile spores. J. Bacteriol., 190(7): 2505–12.

Strathdee, S. A., Hatfull, G. F., Mutalik, V. K. and Schooley, R. T. (2023). Phage therapy: From biological mechanisms to future directions. Cell, 186(1): 17–31.

Tap, J., Mondot, S., Levenez, F., Pelletier, E., Caron, C., Furet, J. P. and Leclerc, M. (2009). Towards the human intestinal microbiota phylogenetic core. Environmental Microbiology, 11(10): 2574–2584.

Theriot, C. M., Koenigsknecht, M. J., Carlson Jr, P. E., Hatton, G. E., Nelson, A. M., Li, B. and Young, V. B. (2014). Antibiotic-induced shifts in the mouse gut microbiome and metabolome increase susceptibility to Clostridium difficile infection. Nat. Commun., 5: 3114.

Tiamani, K., Luo, S., Schulz, S., Xue, J., Costa, R., Khan Mirzaei, M. and Deng, L. (2022). The role of virome in the gastrointestinal tract and beyond. FEMS Microbiol. Rev., 46(6).

Tjellström, B., Stenhammar, L. and Högberg. (2005). Gut microflora associated characteristics in children with celiac disease. Am. J. Gastroenterol., 100(12): 2784–8.

Turnbaugh, P. J., Ley, R. E., Mahowald, M. A., Magrini, V., Mardis, E. R. and Gordon, J. I. (2006). An obesity-associated gut microbiome with increased capacity for energy harvest. Nature, 444(7122): 1027–31.

Vasiliauskas, E. A., Kam, L. Y., Karp, L. C., Gaiennie, J., Yang, H. and Targan, S. R. (2000). Marker antibody expression stratifies Crohn's disease into immunologically homogeneous subgroups with distinct clinical characteristics. Gut, 47(4): 487–96.

Vaziri, N. D., Wong, J., Pahl, M., Piceno, Y. M., Yuan, J., DeSantis, T. Z. and Andersen, G. L. (2013). Chronic kidney disease alters intestinal microbial flora. Kidney Int., 83(2): 308–15.

Vemuri, R., Sylvia, K. E., Klein, S. L., Forster, S. C., Plebanski, M., Eri, R. and Flanagan, K. L. (2019, March). The microgenderome revealed: sex differences in bidirectional interactions between the microbiota, hormones, immunity and disease susceptibility. In Semin. Immunopathol., 41(2): 265–275. Springer Berlin Heidelberg.

Virani, S. S., Alonso, A., Benjamin, E. J., Bittencourt, M. S., Callaway, C. W., Carson, A. P. and American Heart Association Council on Epidemiology and Prevention Statistics Committee and Stroke Statistics Subcommittee. (2020). Heart disease and stroke statistics—2020 update: a report from the American Heart Association. Circulation, 141(9): e139–e596.

Virgin, H. W. (2014). The virome in mammalian physiology and disease. Cell, 157(1): 142–50.

Vogt, N. M., Kerby, R. L., Dill-McFarland, K. A., Harding, S. J., Merluzzi, A. P., Johnson, S. C. and Rey, F. E. (2017). Gut microbiome alterations in Alzheimer's disease. Sci. Rep., 7(1): 13537.

Wang, L. W., Tancredi, D. J. and Thomas, D. W. (2011). The prevalence of gastrointestinal problems in children across the United States with autism spectrum disorders from families with multiple affected members. J. Dev. Behav. Pediatr., 32(5): 351–60.

Wang, T., Cai, G., Qiu, Y., Fei, N., Zhang, M., Pang, X. and Zhao, L. (2012). Structural segregation of gut microbiota between colorectal cancer patients and healthy volunteers. The ISME Journal, 6(2): 320–329.

Willing, B. P., Dicksved, J., Halfvarson, J., Andersson, A. F., Lucio, M., Zheng, Z. and Engstrand, L. (2010). A pyrosequencing study in twins shows that gastrointestinal microbial profiles vary with inflammatory bowel disease phenotypes. Gastroenterology, 139(6): 1844–1854.

Wu, N., Yang, X., Zhang, R., Li, J., Xiao, X., Hu, Y. and Zhu, B. (2013). Dysbiosis signature of fecal microbiota in colorectal cancer patients. Microb. Ecol., 66(2): 462–70.

Wu, S., Morin, P. J., Maouyo, D. and Sears, C. L. (2003). Bacteroides fragilis enterotoxin induces c-Myc expression and cellular proliferation. Gastroenterology, 124(2): 392–400.

Wu, X., Ma, C., Han, L., Nawaz, M., Gao, F., Zhang, X. and Xu, J. (2010). Molecular characterisation of the faecal microbiota in patients with type II diabetes. Curr. Microbiol., 61(1): 69–78.

Yang, Y., Weng, W., Peng, J., Hong, L., Yang, L., Toiyama, Y. and Ma, Y. (2017). Fusobacterium nucleatum increases proliferation of colorectal cancer cells and tumor development in mice by activating toll-like receptor 4 signaling to nuclear factor- κB, and up-regulating expression of microRNA-21. Gastroenterology, 152(4): 851–866.e24.

Yutin, N., Benler, S., Shmakov, S. A., Wolf, Y. I., Tolstoy, I., Rayko, M. and Koonin, E. V. (2021). Analysis of metagenome-assembled viral genomes from the human gut reveals diverse putative CrAss-like phages with unique genomic features. Nat. Commun., 12(1): 1044.

Zárate, S., Taboada, B., Yocupicio-Monroy, M. and Arias, C. F. (2017). Human virome. Arch. Med. Res., 48(8): 701–716.

Zhang, X., Shen, D., Fang, Z., Jie, Z., Qiu, X., Zhang, C. and Ji, L. (2013). Human gut microbiota changes reveal the progression of glucose intolerance. PLoS One, 8(8): e71108.

Zhao, H., Chu, M., Huang, Z., Yang, X., Ran, S., Hu, B., Zhang, C. and Liang, J. (2017). Variations in oral microbiota associated with oral cancer. Scientific Reports, 7(1): 11773. https://doi.org/10.1038/s41598-017-11779-9.

Zuo, T., Liu, Q., Zhang, F., Yeoh, Y. K., Wan, Y., Zhan, H. and Ng, S. C. (2021). Temporal landscape of human gut RNA and DNA virome in SARS-CoV-2 infection and severity. Microbiome, 9(1): 1–16.

Zuo, T., Wu, X., Wen, W. and Lan, P. (2021). Gut microbiome alterations in COVID-19. Genomics Proteomics Bioinformatics, 19(5): 679–688.

CHAPTER 3

The Gut-Brain Axis and the Human Microbiome

Chamma Gupta,[1] *Abhishek Byahut*,[1] *Chandrali Deka*,[1] *Arundhati Bag*[1] and *Bidita Khandelwal*[2,*]

Introduction

The last decade witnessed prolific research on gut microbiota and human health. It is now well-known that gut microbiota (GM) has a positive impact on metabolism, and a disturbed GM can lead to various metabolic disorders, including inflammatory bowel diseases, obesity, type 2 diabetes mellitus (T2DM) and even heart failure. Interestingly, research has found that an altered GM is often associated with neuropsychiatric diseases as well. This can be attributed to the 'gut-brain axis; (GBAx), a link between the enteric microbiome and brain's cognitive centres. This chapter describes human gut microbiota, gut-brain axis and the associated neurological diseases.

[1] Department of Medical Biotechnology, Sikkim Manipal Institute of Medical Sciences, Sikkim Manipal University (SMU), Gangtok, Sikkim, India, 737102.

[2] Department of Medicine, Sikkim Manipal Institute of Medical Sciences, Sikkim Manipal University (SMU), Gangtok, Sikkim, India, 737102.

Emails: chamma28feb@gmail.com; abhishekbyahut93@gmail.com; dekachandrali61@gmail.com; arundhati_bag@rediffmail.com

* Corresponding author: drbidita@gmail.com

Human Microbiome and Gut Microbiota

While 'microbiome' includes all microorganisms and their genetic materials present in different body parts (like skin, mouth, gut and other exposed body parts), the term 'microbiota' indicates organisms in a specific body part, gut microbiota for instance. A decade-long 'Human Microbiome Project' initiated in 2007 by National Institute of Health, USA described healthy human microbiome, which serves as a reference for normal microorganisms residing in the body.

A healthy human gut microbiota alone may contain more than 150 microorganisms (King et al. 2019) and 35,000 species (Jandhyala et al. 2015). Gut bacteria are grouped into three major phyla: Firmicutes, Bacteroidetes and Actinobacteria (Rinninella et al. 2019). Firmicutes and Bacteroidetes phyla constitute 90% of the gut microbiota. The phylum Firmicutes has more than 200 genera, including *Lactobacillus, Bacillus, Enterococcus, Clostridium,* etc., *Clostridium* being the major genera. *Bacteroides* and *Prevotella* are the major genera in phylum Bacteroidetes. Actinobacteria are less abundant and include genus *Bifidobacterium*. Besides the above three phyla, other phyla like Proteobacteria, Fusobacteria and Verrucomicrobia are also present in the gut ecosystem, though occupying only a small part of the GM. Gut Proteobacteria are represented by *Escherichia coli, Helicobacter pylori,* etc.

However, gut microorganisms may vary among individuals. An individual possesses a unique set of GM. This is due to the differential exposure of an infant and an adult to microorganisms – through diet, state of hygiene, stress, etc. A vertical transmission through the placenta, amniotic fluid and meconium initially set the gut microbiome (Clapp et al. 2017). Mode of delivery may also determine an individual's GM. Infants delivered through vagina usually have a higher number of gut microorganisms, compared to those delivered by caesarean section. Breast milk increases intestinal IgA and *Bifidobacterium* species and decreases pro-inflammatory cytokine IL-6, which helps in reducing the risk of gastroenteritis; infants fed with formula instead of breast milk in their first weeks of life have a smaller number of bacterial species (Clapp et al. 2017). An adult gut microbiome predominantly includes *Bacteriodetes* and *Firmicutes* phyla rather than *Bifidobacterium* and *Lactobacillus,* which are more common in infants. Other than bacteria species, yeast and *Candida* species also constitute gut microbiota. Composition of the GM may also change with age.

Gut microorganisms establish a symbiotic relationship with gut mucosa, help in nutrient metabolism and support human health in various ways. For example, they help in the synthesis of some vitamins, like Vitamin B and K, aid in the metabolism of xenobiotics and drugs, and synthesise compounds with antidiabetic and hypolipidemic properties. They contribute to the breakdown of dietary polyphenols for easier absorption. These microorganisms can also form small chain fatty acids (SCFA) during the fermentation of indigestible carbohydrates from diet, in the gut, which helps in maintaining the integrity of gut epithelium and glucose homeostasis, thus helping in lipid metabolism and influencing immune response and inflammation (Morrison and Preston 2016). The microbes also reduce risks of kidney stones (Jandhyala et al. 2015). Current research shows that gut microbiota plays an important role in myelination, neurogenesis and activation as well as maintenance of microglia, hence regulating the gut-brain axis, shaping our behaviour and affecting our mood as well as cognition (Cenit et al. 2017).

The Gut-Brain Axis (GBAx)

This is a two-way signalling mechanism between the gut and the central nervous system, communicating between the intestinal activities and the brain's emotional and cognitive centres (Carabotti et al. 2015). Stimuli from the brain influence the gut's peristaltic movements, and the sensory, secretory and other functions overall. Likewise, signals from the gut also affect brain function. GM has a role in this two-way communication, although the specific mechanism for this is yet to be identified. It is suggested that bacteria may stimulate the enteric nerves and make the signal pass through the vagus nerve. GM may even influence the hypothalamic-pituitary-adrenal axis (HPA) by inducing cytokine production. Conversely, altered HPA function caused due to stress may change GM composition (Figure 1).

The HPA axis is activated by stress factors. Stress stimulates the hypothalamus, which secretes corticotropin hormone (CRH/CRF). CRH, in turn, stimulates the secretion of anterior pituitary adrenocorticotropic hormone (ACTH), which stimulates the adrenal glands to secrete cortisol that responds to stress. Cortisol inhibits CRH and ACTH secretion to maintain homeostasis. HPA is dysregulated in several psychiatric disorders. Although there is limited evidence for humans, strong evidences from rodent studies demonstrate that the composition of GM can influence HPA development. Human studies show that some pre- and probiotics, including some strains of *Lactobacillus*, can decrease salivary cortisol levels (Rosin et al. 2021). Stress may increase gut permeability, allowing bacterial metabolites and bacteria themselves to pass through the submucosa to enter systemic circulation, a condition known as 'leaky gut'. An alteration in the gut's microbial composition, or dysbiosis, may lead to

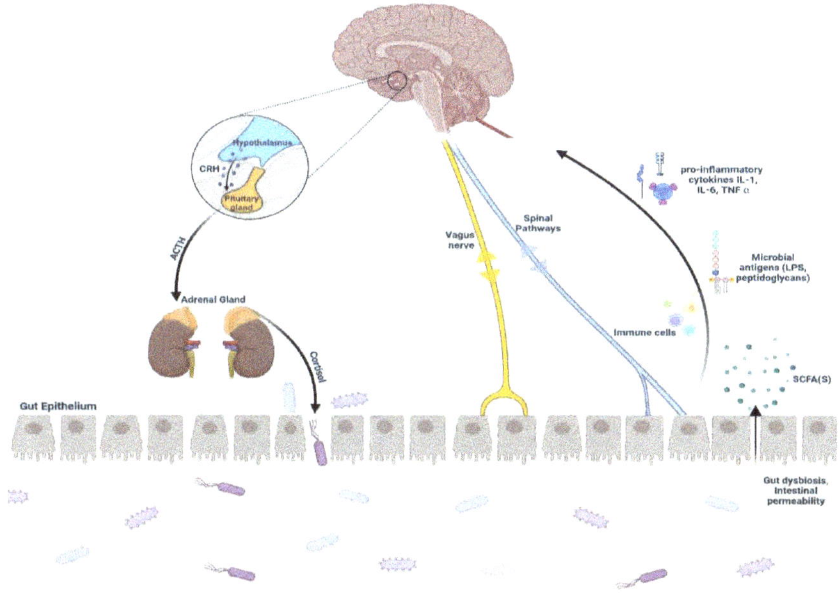

Figure 1. Hypothalamic-Pituitary-Adrenal Axis (HPA).

the enhanced secretion of certain cytokines such as interleukins IL-1β, IL-6 and tumour necrosis factor α (TNF-α), which may cross the blood-brain barrier (BBB) and activate the HPA axis. Furthermore, bacterial cell wall components (lipopoly-saccharides (LPS) and peptidoglycans) can also activate the HPA axis. On the other hand, SCFA can also downregulate genes involved in the HPA axis (Misiak et al. 2020).

GM, through the host's immune system, can also affect functions of the brain – through Toll-like receptors (TLRs) that are expressed abundantly on neurones, for instance (McKernan et al. 2011). Microorganisms of the gut can also produce neurotransmitters like Gamma-aminobutyric acid (GABA), dopamine, serotonin, acetylcholine and nor-epinephrine (Halverson and Alagiakrishnan 2020). Moreover, GM produce brain-derived neurotrophic factor (BDNF), a growth factor which contributes to the growth and survival of neurones, can modulate neurotransmitters, and affects plasticity (Bathina and Das 2015).

Gut-Brain Axis and Diseases

In addition to anatomical, the bidirectional communication of the gut-brain axis connecting the enteric system and the CNS is associated with

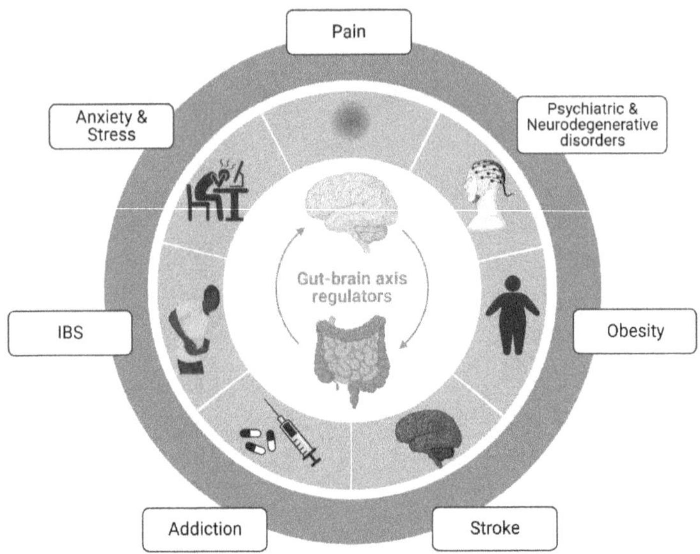

Figure 2. An outline illustrating the diseases associated with the microbiota.

the neurological, endocrinal, humoral, metabolic and immunological pathways for interaction. The autonomic nervous system, the hypothalamic-pituitary-adrenal (HPA) axis, and the nerves within the GI tract – all communicate with the gut and the brain, permitting the brain to regulate GI processes (such as the functioning of immune effector cells) and the gut to affect mood, cognition and mental health. An altered gut-brain axis has often been correlated to psychiatric and neurological disorders like Alzheimer's disease, dementia, Parkinson's disease, Schizophrenia, anxiety, and several others (Figure 2).

1. Gut Brain Axis and neurological disorders
    1.1 Gut Brain Axis and Alzheimer's Disease
    1.2 Gut Brain Axis and Parkinson's Disease
    1.3 Gut Brain Axis and Autism Spectrum Disorder
2. Gut Brain Axis and mental health
    2.1 Gut Brain Axis and mood disorders
    2.2 Gut Brain Axis and anxiety
    2.3 Gut Brain Axis and depression
    2.4 Gut Brain Axis and Schizophrenia

1. Gut Brain Axis and Neurological Disorders

1.1 Gut Brain Axis and Alzheimer's Disease: Alzheimer's disease (AD) is a neurodegenerative brain disorder. It is the most common form of dementia, characterised by a steady loss in cognitive function and involves numerous physiological dysfunctions. Amyloid plaques and neurofibrillary tangles (NFTs) are the two major forms of lesions identified and widely studied in AD. These two lesions are a result of the malfunction and aggregation of two misfolded proteins – the beta-amyloid peptide in amyloid plaques and the tau protein in NFTs. Current research focuses on the intricate biological properties and processes responsible for the aggregation of these two proteins. Model-based studies conducted on animals free of microbes and those subjected to pathological microbial infection, antibiotics, probiotics, or faecal microbiota transplantation indicate the association of the gut microbiota with host cognition or AD-related pathophysiology (Cryan and Dinan 2012).

The disease's hallmark is the build-up of amyloid beta (Aβ), which is followed by the development of plaques and neurofibrillary tangles made of tau protein that has been hyperphosphorylated. These deposits cause neuroinflammation, which leads to synaptic loss and the death of neurons. Although the exact cause of amyloid plaque formation is still unknown, the gut microbiota undoubtedly plays a significant role in the process (Kowalski and Mulak 2019). Tau is a protein that modulates the integrity of axonal microtubules and is very soluble. The tau hypothesis postulates that the aberrant and aggregated forms of this protein may function as toxic triggers that cause neurodegeneration (Jouanne et al. 2017).

***Formation of Amyloid Plaques*:** Amyloid beta is a by-product of the breakdown of the amyloid precursor protein (APP). A transmembrane protein called APP is proteolytically processed by two enzymes called β-secretase and γ-secretase through the amyloidogenic pathway, or α-secretase and γ-secretase through the non-amyloidogenic pathway. APP is cleaved by γ-secretase at the site within the cell membrane and β-secretase at a site outside the cell. Aβ40 and Aβ42 are two misfolded proteins of different lengths, produced by the amyloidogenic pathway. Aβ40 is the most common, while Aβ42 is less prevalent, albeit highly neurotoxic and constituting the basis of the plaque. These peptides serve as seeds that promote the formation of intermediate and hazardous forms of protein oligomers, protofibrils and fibrils accumulated into senile plagues. This system can then self-replicate as the fibril breaks, yielding new seeds (Chen et al. 2017, De Strooper 2010). The most time-consuming stage, the nucleation phase of seed formation is thermodynamically unfavourable and may not take place under physiological circumstances. Exogenous seeds can significantly decrease the period preceding protein aggregation *in vitro* (Jucker and Walker 2013, Kowalski and Mulak 2019).

The microbiota gut-brain axis has the capacity to affect the brain via a network of bidirectional connections comprising neurological, immunological, endocrinal and metabolic pathways. Alzheimer's disease may be profoundly influenced by changes in the brain, gut and microbiome axis. The development of AD is influenced by disturbances along the central nervous system (CNS) and the enteric nervous system (ENS) of the brain-gut-microbiota axis. Due to the production of proinflammatory cytokines, amyloid, lipopolysaccharides and other toxins by pathogenic bacteria, the gut microbiota is known to promote local and systemic inflammation. Bacteria or their by-products may invade the CNS from the oral cavity and the gastrointestinal system, especially in the geriatric population. Bacterial amyloids might cross-seed prion protein misfolding and promote native amyloid aggregation (Jiang et al. 2017). Additionally, the products of the gut microbiota and its modulation may stimulate the microglia, boosting the inflammatory response in the CNS, leading to pathological microglial activity, increased neurotoxicity, increased permeability of the intestinal barrier and the blood-brain barrier, and decreased amyloid clearance. Alteration in the makeup of the gut microbiota can be used as a possible target for therapy in Alzheimer's disease (Jiang et al. 2017).

1.2 **Gut Brain Axis and Parkinson's Disease:** Parkinson's disease (PD) is a multifactorial neurodegenerative disorder, characterised by the build-up and aggregation of a misfolded alpha-synuclein protein (α-syn) known as lewy bodies in the substantia nigra. It is important to produce dopamine in the brain, as well as in other neural structures, the CNS and other tissues. The common symptoms, including bradykinesia (slowness of movement), resting tremor, rigidity and late postural instability, are due to the death of dopamine-producing cells in the substantia nigra (Mulak and Bonaz 2015). A wide range of non-motor symptoms are also present, such as those affecting the gastrointestinal (GI), olfactory (loss of smell), cardiovascular and urogenital systems. It has become evident that PD may influence many levels of the brain-gut axis, including the enteric nervous system (ENS) and the autonomic nervous system

(ANS) (Klann et al. 2022). Despite significant advancements in the field, the exact mechanisms accountable for the initiation and progression of this disease remain unknown. Recent studies reveal a probable connection between commensal gut bacteria and the brain, that might have an impact on neurodevelopment, brain function and overall health. Researchers indicate that non-motor symptoms, such as gastrointestinal manifestations, frequently appear before motor symptoms and the disease diagnosis. This finding supports the idea that the microbiome-gut-brain axis may be involved in the pathological mechanisms underlying PD (Mulak and Bonaz 2015).

***Braak's Hypothesis*:** Braak et al. (2003) and his team hypothesised a possible role of the GI tract in the onset and development of PD. Neurotropic pathogens are the pathogens that cause diseases associated with the CNS. This neurotropic pathogen infiltrates the body via the nasal cavity and is thought to have a potential role in causing aggregation of the α-syn molecules by conformational changes in α-syn molecules. The infection caused can migrate through the gastrointestinal mucosa to the CNS via retrograde transport mechanism. In contrast to other theories, this theory contends that an external pathogen, and not an innate or internally driven mechanism, is more likely to be the probable cause of Parkinson's disease. However, the hypothesis and the associated staging system have historically been questioned for not being applicable to individuals other than a certain subset of those with sporadic Parkinson's disease (Klann et al. 2022).

Healthy intestinal microbiota supports the integrity of the blood-brain barrier by regulating the production of tight junction proteins (occludin and claudin-5) by means of short chain fatty acids (SCFAs). Microbial translocation is known to be linked to local intestinal inflammation, systemic inflammation and neuroinflammation. SCFAs help in maintaining the integrity of the intestinal barrier by inhibiting microbial translocation. However, dysbiosis of the microbiome, associated with an increase in the number of potentially harmful bacteria, can compromise gut barrier integrity because of the endotoxins (lipopolysaccharide) produced by the bacteria that can directly harm intestinal epithelial cells and alter immune responses. The potential link between the GI symptoms of PD and α-syn has been investigated in several studies. While α-syn is widely distributed throughout the brain, it is also found in substantial levels in the ENS. Enteric neurons produce α-syn to mediate the release and uptake of neurotransmitters. The notion that PD pathology may start in the ENS is supported by the evidence of pathogenic α-syn aggregates in the GI tissue biopsies of PD patients. Furthermore, α-syn has been found in the stomach, oesophagus and salivary glands, suggesting its role in typical non-motor symptoms, including hypersalivation, dysphagia, delayed gastric emptying, and gastroparesis.

Access to the complete details about the impact of dietary supplements on the gut microbiota (GM) is crucial for understanding and treating the disease. Live bacteria known as probiotics can aid in managing the health of the host, if administered in right doses. Supplementing the patient's regime with probiotics enhances CNS performance and alleviates the PD's motor and non-motor symptoms (Thangaleela et al. 2022).

Alterations in the composition of the GM and its metabolites has been identified as another key factor contributing to the onset and development of PD. Studies have revealed that the severity of postural instability and gait difficulties (PIGD) in PD patients was related to the *Enterobacteriaceae* content of their GM, indicating a relationship between PIGD and GM. *Prevotellaceae* are less prevalent in the gut, compared to *Lactobacillaceae*; they are associated with a drop in the intestinal hormone ghrelin, which regulates nigrostriatal dopamine. GM may control the production of dopamine by manipulating the dopamine-producing enzymes. *Prevotellaceae* members are less prevalent in PD patients, which might diminish the production of mucin in the gut mucosal layer. Gut permeability is increased by reduced mucin, that aids in an easy access and entry of bacterial toxins and antigens in the body. This facilitates the accumulation of α-syn in the colon and the brain. Inflammation-induced misfolding and accumulation of α-syn can also be caused by reduced butyrate-synthesizing bacteria and increased pro-inflammatory proteobacteria.

Patients suffering from PD have tested positive for the presence of *Clostridium IV, Clostridium XVIII, Holdemania, Aquabacterium, Sphingomonas, Butyricicoccus* and *Anerotruncus*. These organisms correlate with the severity of the disease. *Prevotellaceae, Lachnospiraceae, Lactobacillaceae,* and *Streptococcaceae* are found in lower concentrations in PD patients. In these patients, the relative abundance of the pro-inflammatory bacteria (*Ralstonia*) has been shown to increase, and that of anti-inflammatory butyrate-producing bacteria (*Blautia, Coprococcus, Roseburia,* and *Faecalibacterium*) to decrease. Inflammatory bowel illnesses, GI issues, dysbiosis, immunological dysregulation, and inflammation are the problems observed in PD. Patients with PD frequently experience gastrointestinal issues, such as constipation, and are said to have higher Firmicutes concentrations and lower levels of *Bifidobacterium, Prevotella* and *Lactobacillus*.

The physiological process of ageing causes an imbalance in the pro-inflammatory and anti-inflammatory alterations that have a key role in initiating several human disorders. *Bifidobacteria, Lactobacilli,* and SCFA-producing *Faecalibacterium prausnitzii, Eubacterium* spp., *Roseburia* spp. and *Ruminococcus* spp. are less prevalent in the elderly population. Several signalling pathways involving microbial metabolites may be disrupted and negatively affected by alterations in the microbiota, brought on by ageing preceding gut dysfunction. Thus, aging affects the axial impairment of gait and postural control in PD (Thangaleela et al. 2022).

The gut-brain axis has been established to play a role in the occurrence and progression of many NDs. Studies show that microbial dysbiosis is strongly related to the prevalence of PD. The challenges in comprehending the functional association between GM and PD must be resolved by recognising the microbiological roles in the onset and development of the disease. Altering one's diet can help treat PD that is also accompanied by gut dysbiosis. Understanding the signs of gastrointestinal dysfunction, their connection to Parkinson's disease, and the molecular mechanisms underlying disorders may help develop early detection and treatment techniques.

1.3 **Gut-Brain Axis and Autism Spectrum Disorder:** Autism Spectrum Disorder (ASD) is a highly prevalent neuro-developmental disorder that interferes with normal brain development. ASD is characterised by deficiency in communication, reasoning abilities, repetitive and obstructive behavioural patterns, hyperesthesia and hyperactivity. The multifactorial pathophysiology in the development of ASD includes external factors like dietary intake and stress, as well as genetic factors like chromosomal abnormalities (Mangiola et al. 2016, Matsuzaki et al. 2012). Recent research on the microbiota-gut-brain axis suggests numerous neurological conditions, including autism, to be influenced by gut microbiota. Most autistic people experience GI problems. ASD is caused by an array of factors, including mode of delivery, excessive use of antibiotics, stress, early colonisation on newborn health development, and the impact of prenatal microbiota dysbiosis during the gestation period. These variables ultimately result in the dysbiosis of the gut microbiome and the colonisation of pathogenic bacteria, affecting CNS function by releasing neurotoxins. These pathogenic bacteria, such as *Clostridium*, which are prevalent in children's colon, suggest a risk for ASD development (Taniya et al. 2022).

Different forms of SCFA produced by the microbial fermentation of plant-based fibre may have an advantageous or detrimental effect on the gut and the neurological functioning of autistic patients. Butyrate and propionate are two important SCFA formed by the microbial synthesis of dietary fibre. In contrast to propionate, which has an adverse effect on brain function by making ASD patients more aggressive and prone to behavioural abnormalities, butyrate enhances brain function by blocking histone deacetylases (Yadav et al. 2018).

Following the successful clinical trials on long-term microbial transplant therapy on autistic children, the Food and Drug Administration (FDA) accepted microbial transplant therapy and designated it "fast-track" for ASD treatment in 2019 (Adams et al. 2019).

1.4 Gut-Brain Axis and Dementia: Dementia is a neurocognitive disorder characterised by the gradual impairment of cognitive function. Globally, dementia is an important cause of disabilities among elderly individuals and can impact memory, thinking, language, behaviour, and activities of daily living. Alzheimer's disease (AD) is the most prevalent form of dementia, accounting for 60–80% of all cases. Other types of dementia comprise frontotemporal dementia, Lewy body dementia, Parkinson's disease-related dementia and vascular dementia (Sun et al. 2022).

The gut microbiota has been linked to the onset of dementia. Diagnoses of dementia were made earlier in people with inflammatory bowel ailments connected to the gut microbiota than in healthy samples. In addition, several mental and neurological conditions have the microbiome-gut-brain axis as a possible clinical and therapy target. Modifying gut flora by administering probiotics or antibiotics may improve the efficiency of assessments of learning and memory (Sampson and Mazmanian 2015, Zhang et al. 2021).

2. Gut-Brain Axis and Mental Health: The earliest indication of the gut-brain axis came from the findings of a military surgeon's thorough assessment of gastric secretions released through an intragastric fistula, suggesting a connection between gut health and mood (Mangiola et al. 2016). Irritable bowel syndrome (IBS) is one example of a gastrointestinal disorder that affects about 60% of people with anxiety and depression. Recently, lower microbiome species and genus potent instability have been linked to IBS, along with other changes in the gut microbiota. Meanwhile, several groups of animal studies point to the possibility that gut microbiota abnormalities can lead to an altered brain chemistry and behaviour, as well as an augmentation of the visceral pain response (Park et al. 2013).

According to a reported study (Dickerson et al. 2017), patients with severe psychiatric disorders were found to be prescribed with more antibiotics. It was suggested that an upsurge in the use of antibiotics was associated with the severity of mania; the antibiotics were prescribed mainly against urinary tract infection (UTI) in female patients, while, the prescription was for pulmonary and dermal infections in infected men. The investigators proposed the three potential processes listed below to be present in the recipients of antibiotics:

i. Activation of the immune system during a chronic infection which is consequently followed by mania.
ii. High bacterial infections in mania patients, indicating the poor response state of the immune system.
iii. The administration of antibiotics that may have altered the gut microbiota, which could have increased the likelihood of the mood state.

The diverse human behaviour and mental health problems that may be influenced by the Gut-Brain Axis are mental disorders, anxiety, depression, autism, schizophrenia, etc.

2.1 Gut-Brain Axis and Mood Disorders: Production of metabolites such as bile acids, choline and short-chain fatty acids (SCFA) by the gut flora can help improve the host's metabolism. Gut microbiota are able to break down, absorb and ferment complex carbohydrates like dietary fibres into SCFA comprising butyrate, acetate and propionate. These SCFAs have neuroactive attributes and may circulate in the bloodstream, playing a crucial role in the brain via two different types of dispersed 7-transmembrane G protein coupled receptors such as free fatty acid receptor 2 (FFA2) and FFA3 (Dinan et al. 2015). Thus, the gut-brain axis plays a significant role in mood disorders via gastrointestinal microbiome.

The alteration in the gut microbiota that may be induced by stress can affect mood via the HPA axis. Acute stress is associated with cortisol production via the HPA axis, a biological phenomenon in response to a random stressor, whereas chronic stress is associated with an imbalance of the HPA axis. The gut microbiota is affected differently by these two types of stress. Due to the microbiota's long-term relative stability, acute stress has a limited impact, while chronic stress might disrupt this type of balance. Stress-related alterations in intestinal permeability can be mediated by Corticotropin hormone or factor and CRF/CRH receptors (CRF 1, CRF 2). When under acute stress, visceral hypersensitivity rises along with para-colonic permeability. Stress in the early years of life can raise the level of cortisol in the blood, thereby increasing its permeability and permitting bacteria to make their way to the liver and the spleen (Kelly et al. 2015). The gut microbiota make-up can vary because of stress or mood disorders. In a prenatal stress rat model, maternal stress can lower the amount of *Lactobacillus* in the womb and its transmission to the child, resulting in abnormalities in metabolism associated with energy balance and impacting the amino acid composition in brain development. Higher levels of *Bacillus proteus* (control germs) and comparatively minimal amounts of *Lactobacillus* and *Bifidobacteria* abundance were prevalent in the offspring of mothers who reported elevated levels of stress and saliva cortisol (Petra et al. 2015).

Several types of neurotransmitters regulating the human behaviour can be secreted by the gut microbiota. For instance, *Candida, Streptococcus, Escherichia coli* and *Enterococcus* can release 5-HT, whereas *Bacilli* and *Serratia* can secrete dopamine, and *Lactobacillus* subspecies can secrete acetylcholine (which regulates memory, attention, learning and mood). Some intestinal microbiota, including *Lactobacillus acidophilus, Bifidobacterium infantis, Bifidobacterium, Candida and Streptococcus*, have been shown to have therapeutic effects on mental illnesses by secreting neurotransmitters (GABA, 5-HT, glycine and catecholamine), or by controlling the expression of endocannabinoids. Sleep, hunger, mood and cognition are all neuropsychiatric variables that can be impacted by neuroactive chemicals released by the microbiota of the gut. An alteration in the level of neurotransmitters observed during stress may be induced by the gut microbiota rather than stress (Kali 2016).

Mood disorders are associated with the activity and productive status of neurotrophic chemicals, notably Brain-Derived Neurotrophic Factor (BDNF), suggesting the potential therapeutic applicability of elevated BDNF levels. BDNF is a neurotrophic factor crucial for the development and reparative mechanism of hippocampus and its reduced level helps to balance a rise in 5-HT (Rios et al. 2017). A modest quantity of liposaccharides has been linked to indications of severe anxiety and despair, cognitive impairment and an increase in the sensitivity to visceral pain. Kynurenine released by the gut microbiota or derived directly from food can be promptly absorbed and possesses anti-anxiety attributes; however, kynurenine itself can cause anxiety when operating on the periphery (Mangiola et al. 2016).

Prebiotics, probiotics and appropriate antibiotics are administered as therapeutics for the management of mood disorders. Probiotics help lessen oxidative stress and proinflammatory cytokines in people, both of which are linked to mood disorders. Probiotics work in part by influencing the potential of the gut microbiota, maintaining the integrity of the intestinal barrier, inhibiting the spread of bacteria, and controlling local inflammatory responses via the intestinal immune system (Liu and Zhu 2018).

2.2 **Gut-Brain Axis, Anxiety and Depression:** Several studies show that the activation of stress-induced HPA axis raises cortisone levels, which in turn causes changes in the gut microbiota and elevates anxiety levels. However, gut microbiota can also prevent the increasing release of cortisone by the HPA axis, thereby decreasing anxiety and depression. Depression and anxiety can thus be managed through microbiota. The dearth of intestinal microbiota has been suggested to increase anxiety-like behaviour, when exposed to an open and strong light environment (open field test). These expanding behaviours are accompanied by an aggravated HPA axis response. According to research employing acute open field stress test on gut microbiota-free (GF) rats and special pathogen-free (SPF) rats, higher serum cortisone concentrations have been reported in GF rats than in SPF rats. Antidepressants and *Lactobacillus rhamnosus* are suggested to decrease the rise in cortisone caused by stress (Crumeyrolle-Arias et al. 2014).

Stress can cause alterations in the performance and function of the intestinal barrier, allowing various chemicals to circulate through the blood and the immune system. Elevated levels of proinflammatory IL-1 and IL-6 induced by stress may trigger an inflammatory reaction. According to a study, increased levels of inflammatory biomarkers such IL-6, TNF-α, and C-reactive protein are associated with depression. Recently, depression was reclassified as a clinical manifestation of activated immunological, inflammatory, oxidative and nitrosative stress (IO&NS) routes, including tryptophan metabolism (TRYCAT), autoimmune and gut-brain pathways (Martin-Subero et al. 2016, Petra et al. 2015).

Patients with depression have a significant shift in the gastrointestinal inhabitants of *Acinetobacter* and *Bacteroides*. Reduced levels of

5-HT and BDNF, as well as abnormalities in the configuration of amygdale neurons, are observed in depressed patients, due to an aberrant immune system and brain anomalies. Key by-products of the GM microorganisms include SCFA, butyrate, propionate and acetate. The parasympathetic nervous system is activated by elevated acetate levels caused by a modified gut flora. Hyperphagia, obesity and other associated consequences linked to depression are brought on by a surge in glucose-stimulated insulin and ghrelin (Anadure et al. 2019).

The elevated plasma ratio of kynurenine/tryptophan is indicative of a downregulation of tryptophan metabolism, which is linked to depression. Liposaccharides are crucial for controlling the neural system. They raise the amygdala's activity, which controls emotions, affects the brain's physical activity and controls the continued synthesis of neuropeptides. When administered to healthy individuals, LPS can result in the production of inflammatory cytokines and an increase in plasma norepinephrine, both of which have been associated with greater depressive disorders mortality. Indoleamine 2,3 dioxygenase (IDO), a type of tryptophan-digesting enzyme that acts via the kynurenine pathway, can be activated by LPS-induced inflammation and its activity is strongly associated with depressive symptoms (Mangiola et al. 2016). Prebiotics, probiotics and appropriate antibiotics are offered as therapeutics, regulating the symptoms of depression and anxiety (Liu and Zhu 2018).

2.3 Gut-Brain Axis and Schizophrenia: Schizophrenia (SCZ) is defined as delusions, hallucinations, disorganised speech and behaviour, and other symptoms that result in social or occupational dysfunction (Patrono et al. 2021). The gut microbiota has been shown to be the primary regulator of the gut-brain axis. Over the past ten years, our understanding of the gut microbiome's impact on brain function has grown. Clinical and preclinical investigations have identified the microbiome as a significant contributor to neurological diseases. The gut-brain connection makes it possible for gut microorganisms to communicate with the brain, which could make the microbiota modulators of cognition and behaviour (Vafadari 2021).

Gut microbes and centrally or peripherally mediated behaviour can communicate with one another via the vagus nerve. In addition to neurons, gut microorganisms can control neurotransmitters by altering the levels of their precursors, such as serotonin. Serotonin (5-HT) is a crucial neurotransmitter that mediates communication between the central nervous system and the gastrointestinal tract and is a component of the gut-brain axis. Tryptophan is the precursor to 5-HT, but the brain has limited storage for it. As a result, intestinal refilling is essential, which can be attained by using a bacterium called *Bifidobacterium infantis*. Additionally, some bacteria have the ability to produce and release various neurotransmitters. For instance, *Lactobacillus, Bacillus, Enterococcus, Candida, Streptococcus, Escherichia* and *Saccharomyces* species can generate and release acetylcholine, dopamine, 5-HT and norepinephrine respectively. These interactions are crucial for both sociability and mental wellness. Oral microbiome investigations have shown that individuals with schizophrenia have varying concentrations of Lactobacillus phage phiadh. Different concentrations of Lactobacillus phage phiadh have been associated with the presence of immunological conditions (Vafadari 2021).

The gut-brain axis is found to be significant in psychiatric disorders as per the findings of several animal studies. There is growing evidence that imbalances in the gut microbiome are related to disorders like schizophrenia. It has been demonstrated that the gut microbiome of schizophrenia patients influence the glutamate-glutamine-GABA cycle and schizophrenia-like behaviours in mice models (Zheng et al. 2019). The general microbial make-up of patients with schizophrenia differs significantly from that of non-schizophrenic people (Castro-Nallar et al. 2015). Transplanting the human gut microbiota is reported to improve the treatment of schizophrenia (Fond et al. 2020). The gut microbiota and cognitive dysfunctions resembling SCZ can be used to predict SCZ as a 'wide-spectrum' psychiatric disorder.

3. **Gut-Brain Axis and other diseases:** Several pieces of evidence indicate the implication of gut microbial dysbiosis in the pathophysiology of numerous disorders, including inflammatory bowel disease (IBS), coeliac disease (CD), allergies, asthma, metabolic syndrome, cardiovascular disease, obesity, etc.

Conclusion

In conclusion, the bidirectional communication between gut-brain axis and the human microbiome is a fascinating area of research and has the potential to revolutionise our understanding and treatment of various diseases, including neuropsychiatric and gastrointestinal disorders. The microbiota-gut-brain axis is supported by an expanding array of experimental evidence and clinical observations, and its prospects to regulate fundamental parts of the brain and behaviour in both health and disease.

According to the latest findings, the gut microbiota plays an essential role in the communication between the neurological system and the gut. By controlling brain chemistry and affecting neuroendocrine processes linked to stress response, anxiety, and memory function, it affects the CNS. Stress and excessive consumption of antibiotics have a significant negative impact on the microbiota and can result in disease conditions. The GBA can benefit greatly from prudent antibiotic use and a healthy lifestyle with a balanced diet. Controlled manipulations or modifications of the microbiome is a potential new field of study that may provide treatments for some severe neuro-degenerative conditions, notably Alzheimer's disease, Parkinson's disease, autism, dementia, Schizophrenia and other mental conditions (anxiety, stress, depression). Further research is warranted to fully elucidate the complex interactions between these two systems and to develop more effective interventions for these conditions.

References

Adams, J. B., Borody, T. J., Kang, D. -W., Khoruts, A., Krajmalnik-Brown, R. and Sadowsky, M. J. (2019). Microbiota transplant therapy and autism: Lessons for the clinic. Expert Review of Gastroenterology & Hepatology, 13(11): 1033–1037. https://doi.org/10.1080/17474124.2019.1687293.

Anadure, R., Subramanian, S. and Prasad, A. (2019). The Gut-brain Axis (pp. 1–5).

Bathina, S. and Das, U. N. (2015). Brain-derived neurotrophic factor and its clinical implications. Archives of Medical Science: AMS, 11(6): 1164–1178. https://doi.org/10.5114/aoms.2015.56342.

Braak, H., Del, T. K., Rüb, U., de Vos, R. A., Jansen, S. E. N. and Braak, E. (2003). Staging of brain pathology related to sporadic Parkinson's disease. Neurobiology of Aging, 24(2): 197–211. https://doi.org/10.1016/s0197-4580(02)00065-9.

Carabotti, M., Scirocco, A., Maselli, M. A. and Severi, C. (2015). The gut-brain axis: Interactions between enteric microbiota, central and enteric nervous systems. Annals of Gastroenterology: Quarterly Publication of the Hellenic Society of Gastroenterology, 28(2): 203–209.

Cenit, M. C., Sanz, Y. and Codoñer-Franch, P. (2017). Influence of gut microbiota on neuropsychiatric disorders. World Journal of Gastroenterology, 23(30): 5486–5498. https://doi.org/10.3748/wjg.v23.i30.5486.

Chen, G., Xu, T., Yan, Y., Zhou, Y., Jiang, Y., Melcher, K. and Xu, H. E. (2017). Amyloid beta: Structure, biology and structure-based therapeutic development. Acta Pharmacologica Sinica, 38(9): 1205–1235. https://doi.org/10.1038/aps.2017.28.

Clapp, M., Aurora, N., Herrera, L., Bhatia, M., Wilen, E. and Wakefield, S. (2017). Gut microbiota's effect on mental health: The gut-brain axis. Clinics and Practice, 7(4): 987. https://doi.org/10.4081/cp.2017.987.

Crumeyrolle-Arias, M., Jaglin, M., Bruneau, A., Vancassel, S., Cardona, A., Daugé, V., Naudon, L. and Rabot, S. (2014). Absence of the gut microbiota enhances anxiety-like behavior and neuroendocrine response to acute stress in rats. Psychoneuroendocrinology, 42: 207–217. https://doi.org/10.1016/j.psyneuen.2014.01.014.

Cryan, J. F. and Dinan, T. G. (2012). Mind-altering microorganisms: The impact of the gut microbiota on brain and behaviour. Nature Reviews. Neuroscience, 13(10): 701–712. https://doi.org/10.1038/nrn3346.

De Strooper, B. (2010). Proteases and proteolysis in Alzheimer disease: A multifactorial view on the disease process. Physiological Reviews, 90(2): 465–494. https://doi.org/10.1152/physrev.00023.2009.

Dickerson, F., Severance, E. and Yolken, R. (2017). The microbiome, immunity, and schizophrenia and bipolar disorder. Brain, Behavior, and Immunity, 62: 46–52. https://doi.org/10.1016/j.bbi.2016.12.010.

Dinan, T. G., Stilling, R. M., Stanton, C. and Cryan, J. F. (2015). Collective unconscious: How gut microbes shape human behavior. Journal of Psychiatric Research, 63: 1–9. https://doi.org/10.1016/j.jpsychires.2015.02.021.

Fond, G., Pauly, V., Leone, M., Llorca, P. M., Orleans, V., Loundou, A., Lancon, C., Auquier, P., Baumstarck, K. and Boyer, L. (2021). Disparities in Intensive Care Unit Admission and Mortality Among Patients With Schizophrenia and COVID-19: A National Cohort Study. Schizophrenia Bulletin, 47(3): 624–634. https://doi.org/10.1093/schbul/sbaa158.

Halverson, T. and Alagiakrishnan, K. (2020). Gut microbes in neurocognitive and mental health disorders. Annals of Medicine, 52(8): 423–443. https://doi.org/10.1080/07853890.2020.1808239.

Jandhyala, S. M., Talukdar, R., Subramanyam, C., Vuyyuru, H., Sasikala, M. and Reddy, D. N. (2015). Role of the normal gut microbiota. World Journal of Gastroenterology: WJG, 21(29): 8787–8803. https://doi.org/10.3748/wjg.v21.i29.8787.

Jiang, C., Li, G., Huang, P., Liu, Z. and Zhao, B. (2017). The gut microbiota and Alzheimer's disease. Journal of Alzheimer's Disease: JAD, 58(1): 1–15. https://doi.org/10.3233/JAD-161141.

Jouanne, M., Rault, S. and Voisin-Chiret, A. -S. (2017). Tau protein aggregation in Alzheimer's disease: An attractive target for the development of novel therapeutic agents. European Journal of Medicinal Chemistry, 139: 153–167. https://doi.org/10.1016/j.ejmech.2017.07.070.

Jucker, M. and Walker, L. C. (2013). Self-propagation of pathogenic protein aggregates in neurodegenerative diseases. Nature, 501(7465): 45–51. https://doi.org/10.1038/nature12481.

Kali, A. (2016). Psychobiotics: An emerging probiotic in psychiatric practice. Biomedical Journal, 39(3): 223–224. https://doi.org/10.1016/j.bj.2015.11.004.

Kelly, J., Kennedy, P., Cryan, J., Dinan, T., Clarke, G. and Hyland, N. (2015). Breaking down the barriers: The gut microbiome, intestinal permeability and stress-related psychiatric disorders. Frontiers in Cellular Neuroscience, 9. https://www.frontiersin.org/articles/10.3389/fncel.2015.00392.

King, C. H., Desai, H., Sylvetsky, A. C., LoTempio, J., Ayanyan, S., Carrie, J., Crandall, K. A., Fochtman, B. C., Gasparyan, L., Gulzar, N., Howell, P., Issa, N., Krampis, K., Mishra, L., Morizono, H., Pisegna, J. R., Rao, S., Ren, Y., Simonyan, V., … Mazumder, R. (2019). Baseline human gut microbiota profile in healthy people and standard reporting template. PloS One, 14(9): e0206484. https://doi.org/10.1371/journal.pone.0206484.

Klann, E. M., Dissanayake, U., Gurrala, A., Farrer, M., Shukla, A. W., Ramirez-Zamora, A., Mai, V. and Vedam-Mai, V. (2022). The Gut–Brain Axis and its relation to Parkinson's disease: A review. Frontiers in Aging Neuroscience, 13: 782082. https://doi.org/10.3389/fnagi.2021.782082.

Kowalski, K. and Mulak, A. (2019). Brain-Gut-Microbiota Axis in Alzheimer's disease. Journal of Neurogastroenterology and Motility, 25(1): 48–60. https://doi.org/10.5056/jnm18087.

Liu, L. and Zhu, G. (2018). Gut–Brain Axis and mood disorder. Frontiers in Psychiatry, 9. https://www.frontiersin.org/articles/10.3389/fpsyt.2018.00223.

Mangiola, F., Ianiro, G., Franceschi, F., Fagiuoli, S., Gasbarrini, G. and Gasbarrini, A. (2016). Gut microbiota in autism and mood disorders. World Journal of Gastroenterology, 22(1): 361–368. https://doi.org/10.3748/wjg.v22.i1.361.

Martin-Subero, M., Anderson, G., Kanchanatawan, B., Berk, M. and Maes, M. (2016). Comorbidity between depression and inflammatory bowel disease explained by immune-inflammatory, oxidative, and nitrosative stress; tryptophan catabolite; and gut–brain pathways. CNS Spectrums, 21(2): 184–198. https://doi.org/10.1017/S1092852915000449.

Matsuzaki, H., Iwata, K., Manabe, T. and Mori, N. (2012). Triggers for Autism: Genetic and environmental factors. Journal of Central Nervous System Disease, 4: JCNSD.S9058. https://doi.org/10.4137/JCNSD.S9058.

McKernan, D. P., Dennison, U., Gaszner, G., Cryan, J. F. and Dinan, T. G. (2011). Enhanced peripheral toll-like receptor responses in psychosis: Further evidence of a pro-inflammatory phenotype. Translational Psychiatry, 1(8): e36. https://doi.org/10.1038/tp.2011.37.

Misiak, B., Łoniewski, I., Marlicz, W., Frydecka, D., Szulc, A., Rudzki, L. and Samochowiec, J. (2020). The HPA axis dysregulation in severe mental illness: Can we shift the blame to gut microbiota? Progress in Neuro-Psychopharmacology & Biological Psychiatry, 102: 109951. https://doi.org/10.1016/j.pnpbp.2020.109951.

Morrison, D. J. and Preston, T. (2016). Formation of short chain fatty acids by the gut microbiota and their impact on human metabolism. Gut Microbes, 7(3): 189–200. https://doi.org/10.1080/19490976.2015.1134082.

Mulak, A. and Bonaz, B. (2015). Brain-gut-microbiota axis in Parkinson's disease. World Journal of Gastroenterology: WJG, 21(37): 10609–10620. https://doi.org/10.3748/wjg.v21.i37.10609.

Park, A. J., Collins, J., Blennerhassett, P. A., Ghia, J. E., Verdu, E. F., Bercik, P. and Collins, S. M. (2013). Altered colonic function and microbiota profile in a mouse model of chronic depression. Neurogastroenterology & Motility, 25(9): 733–e575. https://doi.org/10.1111/nmo.12153.

Patrono, E., Svoboda, J. and Stuchlík, A. (2021). Schizophrenia, the gut microbiota, and new opportunities from optogenetic manipulations of the gut-brain axis. Behavioral and Brain Functions: BBF, 17(1): 7. https://doi.org/10.1186/s12993-021-00180-2.

Petra, A. I., Panagiotidou, S., Hatziagelaki, E., Stewart, J. M., Conti, P. and Theoharides, T. C. (2015). Gut-Microbiota-Brain Axis and its effect on neuropsychiatric disorders with suspected immune dysregulation. Clinical Therapeutics, 37(5): 984–995. https://doi.org/10.1016/j.clinthera.2015.04.002.

Rinninella, E., Raoul, P., Cintoni, M., Franceschi, F., Miggiano, G. A. D., Gasbarrini, A. and Mele, M. C. (2019). What is the healthy gut microbiota composition? A changing ecosystem across age, environment, diet, and diseases. Microorganisms, 7(1): 14. https://doi.org/10.3390/microorganisms7010014.

Rios, A. C., Maurya, P. K., Pedrini, M., Zeni-Graiff, M., Asevedo, E., Mansur, R. B., Wieck, A., Grassi-Oliveira, R., McIntyre, R. S., Hayashi, M. A. F. and Brietzke, E. (2017). Microbiota abnormalities and the therapeutic potential of probiotics in the treatment of mood disorders. Reviews in the Neurosciences, 28(7): 739–749. https://doi.org/10.1515/revneuro-2017-0001.

Rosin, S., Xia, K., Azcarate-Peril, M. A., Carlson, A. L., Propper, C. B., Thompson, A. L., Grewen, K. and Knickmeyer, R. C. (2021). A preliminary study of gut microbiome variation and HPA Axis reactivity in healthy infants. Psychoneuroendocrinology, 124: 105046. https://doi.org/10.1016/j.psyneuen.2020.105046.

Sampson, T. R. and Mazmanian, S. K. (2015). Control of brain development, function, and behavior by the microbiome. Cell Host & Microbe, 17(5): 565–576. https://doi.org/10.1016/j.chom.2015.04.011.

Sun, H. -L., Feng, Y., Zhang, Q., Li, J. -X., Wang, Y. -Y., Su, Z., Cheung, T., Jackson, T., Sha, S. and Xiang, Y. -T. (2022). The Microbiome–Gut–Brain Axis and Dementia: A bibliometric analysis. International Journal of Environmental Research and Public Health, 19(24): Article 24. https://doi.org/10.3390/ijerph192416549.

Taniya, M. A., Chung, H. -J., Al Mamun, A., Alam, S., Aziz, Md. A., Emon, N. U., Islam, Md. M., Hong, S. -T. shool, Podder, B. R., Ara Mimi, A., Aktar Suchi, S. and Xiao, J. (2022). Role of gut microbiome in autism spectrum disorder and its therapeutic regulation. Frontiers in Cellular and Infection Microbiology, 12. https://www.frontiersin.org/articles/10.3389/fcimb.2022.915701.

Thangaleela, S., Sivamaruthi, B. S., Kesika, P., Bharathi, M. and Chaiyasut, C. (2022). Role of the Gut–Brain Axis, gut microbial composition, diet, and probiotic intervention in Parkinson's disease. Microorganisms, 10(8): Article 8. https://doi.org/10.3390/microorganisms10081544.

Vafadari, B. (2021). Stress and the role of the Gut-Brain Axis in the pathogenesis of Schizophrenia: A literature review. International Journal of Molecular Sciences, 22(18): 9747. https://doi.org/10.3390/ijms22189747.

Yadav, M., Verma, M. K. and Chauhan, N. S. (2018). A review of metabolic potential of human gut microbiome in human nutrition. Archives of Microbiology, 200(2): 203–217. https://doi.org/10.1007/s00203-017-1459-x.

Zhang, B., Wang, H. E., Bai, Y. -M., Tsai, S. -J., Su, T. -P., Chen, T. -J., Wang, Y. -P. and Chen, M. -H. (2021). Inflammatory bowel disease is associated with higher dementia risk: A nationwide longitudinal study. Gut, 70(1): 85–91. https://doi.org/10.1136/gutjnl-2020-320789.

Chapter 4

Diabetes Mellitus and the Microbiome

Helena P. Felgueiras

Introduction

Diabetes mellitus (DM) is one of the most important diseases of the 21st century, raising serious social-economical concerns. According to the World Health Organization (WHO), DM is labelled as a chronic, metabolic disease characterized by hyperglycemia and affects more than 422 million people worldwide, with rates continuously increasing each year and the majority of the affected belonging to low- to middle-income countries (WHO 2023). Knowing DM to be a leading cause of cardiovascular diseases, renal failure, and blindness, it is no surprise that high mortality rates are also associated with this disease (Zhang et al. 2021). In fact, WHO reports that an estimated 1.5 million deaths per year are directly linked to DM and predicts that DM will be the seventh primary cause of death in the world by 2030 (Mathers and Loncar 2006, WHO 2023). Because of the health constraints introduced by DM in the affected people, namely impaired immune response and high microbial counts, alterations in the human microbiome occur, which becomes compromised and aggravates the disease (Zhang et al. 2021).

Centre for Textile Science and Technology (2C2T), University of Minho, Campus de Azurém, 4800-058 Guimarães, Portugal.
Email: helena.felgueiras@2c2t.uminho.pt

The human microbiome refers to the collection of microorganisms that live on and inside the human body. These microorganisms (bacteria, viruses and fungi) are found in various parts of the body, such as the skin, gut and oral cavity. They play important roles in maintaining the overall health of an individual by aiding in digestion, regulating the immune system, and preventing the growth of harmful bacteria (Botero et al. 2016, Komaroff 2017). Additionally, the complex ecosystem that is the microbiota is responsible for over 98% of the genetic activity of the organism, is a source of genetic diversity, and is an entity capable of great influence over metabolism and the interactions that many drugs establish with our system (Grice and Segre 2012).

DM patients have shown alterations in their microbiota, compared to non-diabetic individuals. Specifically, diabetic patients tend to have a lower diversity of gut bacteria and an overgrowth of some potentially harmful microbials. This shift in the microbial community can contribute to inflammation and insulin resistance (Gomes et al. 2014, Wu et al. 2020). In the skin microbiome, diabetes can affect its ability to maintain a healthy microbial environment, which can lead to an overgrowth of certain types of bacteria and fungi and consequently contribute to skin infections, such as diabetic foot ulcers (which are very difficult to treat and may lead to serious complications, including limb loss by amputation) (Jneid et al. 2017). In the oral microbiota, alterations may also be observed with the development of periodontal diseases and other oral health issues which are triggered by the progression of diabetes and may exert some influence on the efficacy of diabetic treatments (Long et al. 2017). In the present chapter, we explore this relationship further by analyzing the mutual influence that DM and the human microbiome exert on each other, and highlight their roles in a patient's immune response. Particular attention has been paid to recent studies on skin, gut and oral microbiomes of DM patients.

Diabetes Mellitus

DM is a physiological dysfunction resulting from insulin resistance or excessive glucagon secretion and can be generally categorized into Type 1 and Type 2 diabetes. In a Type 1 scenario, patients are afflicted with an autoimmune disorder that targets pancreatic β-cells, whereas in Type 2 diabetes, the hyperglycemia problem is aggravated over time by glucose dysregulation (Blair 2016). Type 2 diabetes is more frequent in people with known risk factors (i.e., obesity) and genetic predisposition or may occur in response to environmental triggers (i.e., viruses) (Figure 1) (Fang et al. 2019). Obesity is perhaps the most important effector in Type 2 diabetes

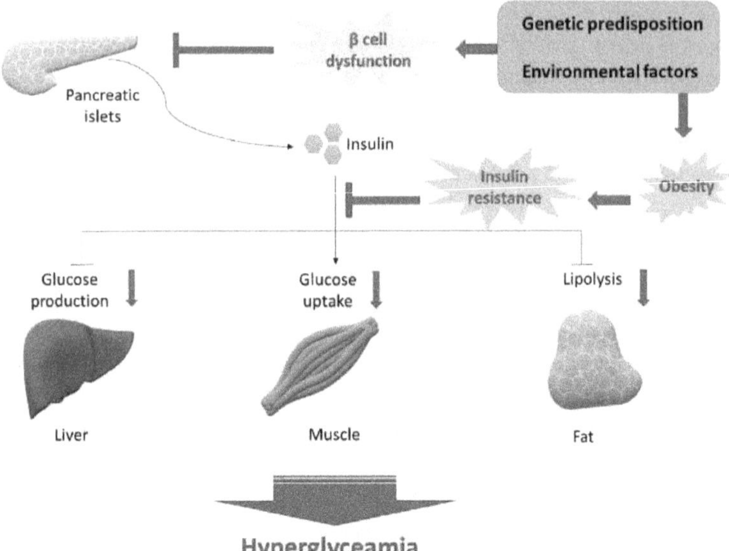

Figure 1. Type 2 diabetes pathology: β-cell dysfunction and insulin resistance occurs in response to genetic predisposition and environmental factors (CC BY 4.0 permission from Fang et al. (2019)).

development, considering that it is often linked to high fat/carbohydrate intakes and a high release of adipokines and cytokines in turn. These substances trigger inflammation and lead to a state of chronic low-grade inflammation, which can continuously interfere with the body's insulin signaling pathways and, over time, contribute to insulin resistance (Kahn and Flier 2000).

The main symptoms associated with DM include hyperglycemia over extended time (high glucose levels), regular urination, augmented thirst and elevated hunger. Early diagnosis, via routine biochemical tests like glycosylated hemoglobin and oral glucose tolerance tests, may prevent serious health problems from developing. Yet, as the disease evolves, so do the symptoms and their complications. Diabetic nephropathy and retinopathy are some of the complications that arise from long-term exposure to DM. Diabetic nephropathy refers to kidney damage and is a common problem in both Type 1 and Type 2 diabetes. Injury in the basement of glomerular capillaries, disruption of protein crosslinking, and passage of urine proteins to the kidneys are major effectors in this condition. They can lead to the progressive loss of kidney function, eventually resulting in kidney failure if left untreated (Lim 2014). On its turn, diabetic retinopathy is a complication that affects the eyes. It occurs when high blood sugar levels cause damage to the blood vessels in the retina. Over time, diabetic retinopathy can lead to vision loss or even blindness (Tarr et al. 2013). Another common complication, frequently detected in Type 2 diabetic patients, is ketoacidosis. Diabetic ketoacidosis is related to insulin insufficiency and results from the continuous production of ketone bodies. Additionally, DM patients are at great risk of developing coronary artery diseases and heart diseases, and sudden cardiac death. DM has also been associated with cognitive dysfunctions that affect attention, memory, learning and perception. However, most importantly, DM has been highlighted as a risk factor in many cancers since the complications behind DM are the same as those behind cancer (Alam et al. 2021).

The complex nature of diabetes and its variability among patients hinders the development of effective therapies universal to all. Even though a cure is highly sought after, the etiology of diabetes makes it extremely challenging. As such, controlled nutrient intake by means of diet restrictions, physical exercise, lipid ingestion monitoring, and use of appropriate medication (i.e., insulin injections) – all under the supervision of health professionals – remain the standard management practices for DM patients. Medication is often the last resort in managing diabetes, with 100% of Type 1 and 25–30% of Type 2 diabetic patients requiring insulin intake (Alam et al. 2021, Tisch and McDevitt 1996).

Impact on the Skin Microbiome

The skin is a home to a complex microbial community of bacteria, fungi, viruses and mites that play a critical role in maintaining an individual's health. These microorganisms interact with each other and with the host, sustaining a delicate balance. They are essential for protecting the organism against invading pathogens, regulating, prompting an effect or educating our immune system, and breaking down natural products into their basic units for facilitated elimination or adsorption (Byrd et al. 2018). Depending on the region of the body, the skin microbiome can present a great diversity. Indeed, the armpits, groin and feet are the richest in microorganisms because of the accumulated sweat and existing sebaceous glands – feeding sources of metabolites for these microorganisms (Figure 2). These microbials have adapted to survive and thrive in the low-nutrient environment that characterize the skin, taking advantage, for instance, of the lipase-mediated degradation of sebum lipids and of the enzymatic cleavage of sphingomyelin for adhering, multiplying and colonizing these regions of the skin (Schommer and Gallo 2013, Swaney et al. 2023).

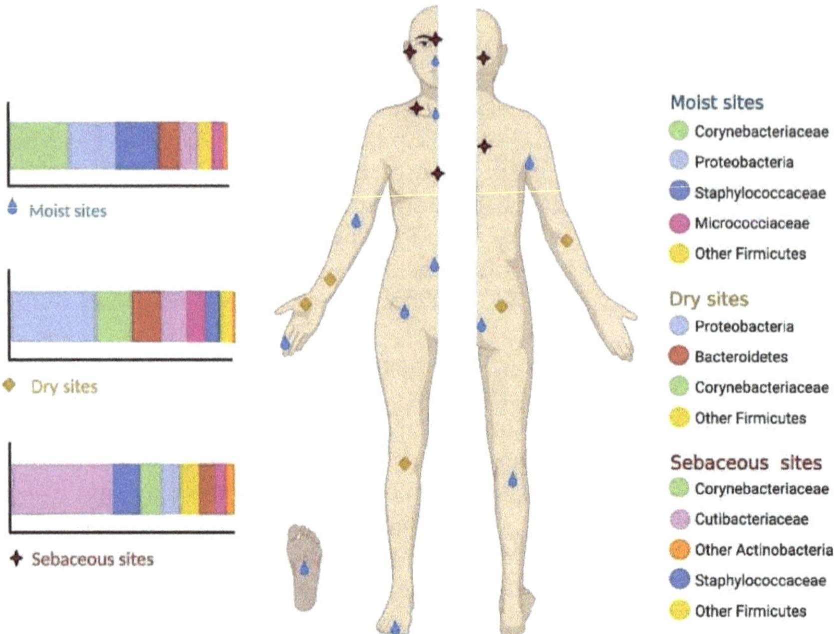

Figure 2. Skin microbiome: main composition and preferential location among moist, dry and sebaceous sites (CC BY permission from Carmona-Cruz et al. (2022)).

Emerging evidence suggests that the skin microbiome can play a crucial role in the development and progression of diabetes. Indeed, when the immune system is impaired, and considering that every square centimeter of the surface of the skin contains about one billion bacteria, DM patients are at an increased risk of developing diabetic foot ulcers (DFUs) if they suffer a skin injury (Grice et al. 2008). Several studies have demonstrated that there is a significant difference between the composition of the microbiomes of diabetic and healthy individuals, with DM patients having higher populations of *Staphylococcus aureus* bacterium, along with phylum Actinobacteria from the genus Corynebacterium (Redel et al. 2013). Additionally, reports have shown that the skin microbiome of a diabetic patient is less diverse than that of a healthy individual, considering dysfunctions in the sebaceous and sweat glands together with the inhibition of many innate antimicrobial peptides, specifically towards *S. aureus* bacterium, altering the microenvironment and affecting its population (Redel et al. 2013, Schittek et al. 2001). These imbalances in composition and numbers, also known as dysbiosis, are responsible for instigating inflammatory events that put the patient at a risk of developing infections and, consequently, chronic wounds (Zhang et al. 2021). Dysbiosis of the skin microbiome in DM patients is associated with increased counts of inflammatory cells, raising dermal infiltration (facilitating microbial access and inflammatory cues migration towards the systemic circulation) and exacerbating the expression of the matrix metallopeptidase 9 and the protein tyrosine phosphatase-1B, ultimately increasing resistance to the action of growth factors, thus hindering wound healing. This occurs because, as the microbiota infiltrates a wound, the microorganisms' signaling pathways, which are similar to those of pathogenic agents, are recognized as belonging to pathogens and trigger an endless inflammatory cascade (Dinh et al. 2012, Zhang et al. 2021). The cutaneous inflammatory cues that reach the systemic circulation release proinflammatory cytokines, triggering a chronic systemic inflammatory state and metabolic syndrome that is regarded as another underlying connection between the skin microbiome and DM (Nakatsuji et al. 2016). Additionally, exposure to large amounts of *S. aureus* bacterium may also disrupt the immune system, compromising glucose tolerance, and thus generate an imbalance in pro- and anti-inflammatory immune cells that lead to insulin resistance (Zhang et al. 2021). Even though evidence has been found regarding the mechanisms underlying the relationship between the skin microbiome and diabetes, there is still much that remains to be understood.

Impact on the Gut Microbiome

The gut microbiome is made of a vast community of microorganisms that inhabit the human gastrointestinal tract. The most abundant bacteria are from the phyla *Firmicutes, Actinobacteria* and *Bacteroidetes*, with the genus *Bacteroides* comprising more than 30% of those. These microorganisms play important roles in digestion, nutrient absorption, hormone modulation, synthesis of vitamins and amino acids, and immune protection, to name a few. Microbial diversity is key to a healthy gut microbiome, considering different microbes can conduct similar functions. Disruptions in the gut microbiome by means of dysbiosis, or imbalanced microbial population levels, have been linked to a range of conditions, including inflammatory bowel disease, obesity, and metabolic disorders such as diabetes (Figure 3) (Singer-Englar et al. 2019).

In DM patients, alterations in the gut microbiota have been associated with changes in insulin sensitivity, glucose metabolism, and the development of metabolic syndromes. It has been reported that microbial diversity tends to decrease in Type 2 diabetic scenarios, with a significant reduction of *Clostridium, Fecalibacteria* and *Roseburia* bacteria. These bacteria have been shown to upregulate butyrate, which can improve glucose levels, reduce insulin resistance, and decrease inflammatory markers (Jia

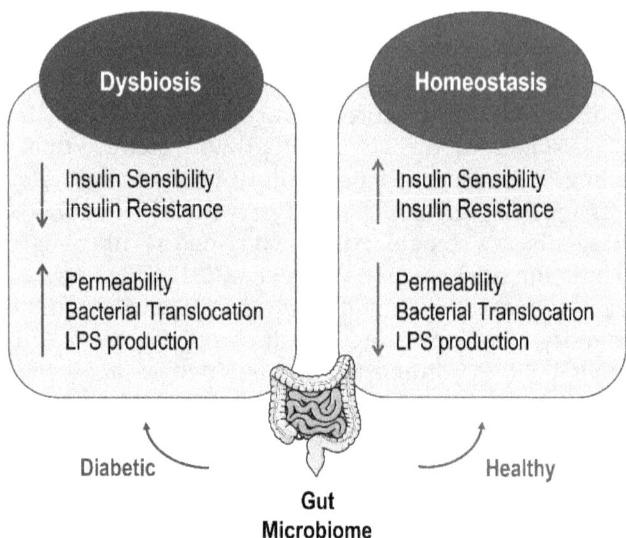

Figure 3. Gut microbiota's influence on the progression and prevention of diabetes in dysbiosis and homeostasis states.

et al. 2017). In fact, different pathways analyses have determined that Type 2 diabetic patients with low amounts of these butyrate-producing bacteria present microbial environments enriched with opportunistic pathogens that led to an increased membrane sugar presence as well as branched-chain amino acid transport, sulfate reduction and augmented oxidative stress resistance, hence confirming the protective features of these microbials (Hur and Lee 2015). Additionally, Allin et al. (2018) revealed that a reduction of the mucin-degrading bacterium *Akkermansia muciniphila* in pre-diabetic patients weakens glucose tolerance, increases insulin resistance and augments adipose tissue inflammation (Allin et al. 2018).

In patients diagnosed with the autoimmune Type 1 diabetes, the pro-inflammatory gut environment is governed by a relative abundance of *Bacteroidetes* and decreased presence of *Firmicutes,* like *Clostridium* strains. Short-chain fatty acids (SCFAs), like butyrate, propionate and acetate, are among the metabolites that originated from these bacteria. As a consequence of the reduced populations of *Firmicutes,* low levels of SCFAs are generated, which exacerbate the host's autoimmune response (Siljander et al. 2019, Zhang et al. 2021). SCFAs can influence the parasympathetic activity to support glucose-stimulated insulin secretion, aside from strengthening epithelial barrier functions and boosting antibody production and macrophage- and dendritic-based regulation of T cells, for a superior protection against pathogens (Perry et al. 2016,

Zhang et al. 2021). Furthermore, the reduction in SCFAs observed in individuals with diabetes has been linked to a decrease in the production of gut hormones GLP-1 (glucagon-like peptide-1) and PYY (peptide YY). GLP-1 is known to stimulate insulin secretion from pancreatic β-cells in response to rising blood glucose levels, and inhibit glucagon secretion from pancreatic α-cells, thus preventing the liver from releasing glucose into the bloodstream. On its turn, PYY acts on the hypothalamus of the brain to reduce appetite and can also stimulate insulin secretion from pancreatic β-cells, although its effect is less pronounced than that of GLP-1 (Stanley et al. 2004).

In both Type 1 and Type 2 diabetes, the gut microbiome is rich in lipopolysaccharides (LPS)-producing bacteria, namely *Bacteroidetes*. These LPS can trigger an immune response by binding to the cluster of differentiation 14 (CD14), a protein found on the surface of certain immune cells like monocytes and macrophages, leading to chronic inflammation via cytokine production and chemokine-mediated recruitment of inflammatory cells. These events are particularly noticeable in Type 2 diabetic patients as they display an augmented response to oxidative stress, which can be correlated with a prolonged pro-inflammatory state (Shi et al. 2006). In recent years, significant efforts have been made to modify the gut microbiome composition through dietary interventions or probiotic supplements, which may reduce the abundance of LPS-producing bacteria and assist in the maintenance of a healthy gut microbiota (Salgaço et al. 2019).

Impact on the Oral Microbiome

The oral cavity is colonized by a diverse array of microorganisms, collectively known as the oral microbiome. These microorganisms play a critical role in maintaining oral health by participating in various physiological processes, such as the digestion of food, synthesis of vitamins, and protection against pathogens. However, alterations in the composition and function of the oral microbiome may lead to the development of various oral diseases, including dental caries, periodontitis and oral cancer. Among the 700+ species isolated from the oral microbiome (57% already identified), pathogens like *Streptococcus mutans, Porphyromonas gingivalis, Tannerella forsythia* and *Aggregatibacter actinomycetemcomitans* can be highlighted as the ones behind the emergence of the referred oral problematics (Sedghi et al. 2021). Additionally, as the oral microbiome is constantly exposed to exogenous foreign substances, colonization by other pathogens capable of resisting the oral cavity microenvironment and the amylase enzyme action may occur, augmenting the local microbial population (Baker and Edlund 2019).

The relationship between oral microbiome and DM is not yet fully understood. In fact, very few conclusions have been drawn from the research done on the subject because of the variability in microbiota composition between patients. In 1993, Löe (1993) reported that Type 1 and Type 2 diabetic patients experienced a three- to four-fold increased risk of periodontitis, a chronic bacterial infection that affects the gingiva and the bone that supports the teeth (one of the major complications of DM) (Löe 1993). It was later determined that periodontitis sites in Type 1 diabetic patients were colonized mainly by *P. gingivalis*, *A. actinomycetemcomitans* and *Campylobacter* spp. (Ebersole et al. 2008), while *P. gingivalis* and *Candida* spp. dominated in Type 2 diagnosed patients (Sardi et al. 2011). Also, Casarin et al. (2013) demonstrated a rise in species like *Veillonella parvula*, *Veillonella dispar*, *Fusobacterium nucleatum* and *Eikenella corrodens* in DM-affected people (Casarin et al. 2013). More recently, studies have verified the dependency between glycemic statuses and different stages of periodontal disease on the alterations of oral microbiome composition and number (dysbiosis) (Matsha et al. 2020). However, to date, it remains unclear if these changes in the microbiome are a result of the hyperglycemic state or the presence of periodontal diseases; many assume that a bidirectional influence is in place (Negrato et al. 2013).

Aside from the microbiome alterations in composition and number, translocation of oral bacteria into the systemic circulation has been shown to activate immune cells and release pro-inflammatory cytokines such as interleukin-1 beta (IL-1β) and tumor necrosis factor-alpha (TNF-α), leading to a chronic inflammatory state and impaired glucose metabolism. The oral microbiome can also influence the production of anti-inflammatory cytokines, such as interleukin-10 (IL-10) and transforming growth factor-beta (TGF-β), which can counteract the pro-inflammatory effects and protect against insulin resistance. The complex interactions between cytokines that feedback in the pro- and anti-inflammatory circuitry determine the development of insulin resistance (Negrini et al. 2021).

Conclusion

The human microbiome is a complex and dynamic ecosystem that plays a critical role in maintaining the organism's health. Alterations in the microbiome composition, number and function have been implicated in the pathogenesis of diabetes, a chronic metabolic disorder that affects millions of people worldwide. As witnessed, the skin, gut and oral microbiome are all important microbial communities implicated in this disease. Despite the information presented in this chapter, there is still much to uncover regarding the mutual influence between the human microbiome and DM. In fact, to this date, there is no consensus on the part which exerts the influence and triggers the response. For instance, many authors suggest that dysbiosis is an effect of DM, while others defend that the converse is true. Additional data is still required to comprehend the local and systemic effects that the local and the infiltrated microorganisms from the microbiome exert on the progression of DM.

Acknowledgements

This research was funded by the Portuguese Foundation for Science and Technology (FCT), via grant UIDP/00264/2020 of 2C2T Strategic Project 2020–2023. H.P.F. also acknowledges FCT for contract 2021.02720. CEECIND.

References

Alam, S., Hasan, M. K., Neaz, S., Hussain, N., Hossain, M. F. and Rahman, T. (2021). Diabetes mellitus: Insights from epidemiology, biochemistry, risk factors, diagnosis, complications and comprehensive management. Diabetology, 2(2): 36–50. https://www.mdpi.com/2673-4540/2/2/4.

Allin, K. H., Tremaroli, V., Caesar, R., Jensen, B. A., Damgaard, M. T., Bahl, M. I., Licht, T. R., Hansen, T. H., Nielsen, T. and Dantoft, T. M. (2018). Aberrant intestinal microbiota in individuals with prediabetes. Diabetologia, 61: 810–820.

Baker, J. L. and Edlund, A. (2019). Exploiting the oral microbiome to prevent tooth decay: Has evolution already provided the best tools? Frontiers in Microbiology, 9: 3323.

Blair, M. (2016). Diabetes mellitus review. Urologic Nursing, 36(1).

Botero, L. E., Delgado-Serrano, L., Cepeda Hernandez, M. L., Del Portillo Obando, P. and Zambrano Eder, M. M. (2016). The human microbiota: The role of microbial communities in health and disease. Acta Biológica Colombiana, 21(1): 5–15.

Byrd, A. L., Belkaid, Y. and Segre, J. A. (2018). The human skin microbiome. Nature Reviews Microbiology, 16(3): 143–155.

Carmona-Cruz, S., Orozco-Covarrubias, L. and Sáez-de-Ocariz, M. (2022). The human skin microbiome in selected cutaneous diseases. Frontiers in Cellular and Infection Microbiology, 145.

Casarin, R., Barbagallo, A., Meulman, T., Santos, V., Sallum, E., Nociti, F., Duarte, P., Casati, M. and Gonçalves, R. (2013). Subgingival biodiversity in subjects with uncontrolled type-2 diabetes and chronic periodontitis. Journal of Periodontal Research, 48(1): 30–36.

Dinh, T., Tecilazich, F., Kafanas, A., Doupis, J., Gnardellis, C., Leal, E., Tellechea, A., Pradhan, L., Lyons, T. E. and Giurini, J. M. (2012). Mechanisms involved in the development and healing of diabetic foot ulceration. Diabetes, 61(11): 2937–2947.

Ebersole, J. L., Holt, S. C., Hansard, R. and Novak, M. J. (2008). Microbiologic and immunologic characteristics of periodontal disease in Hispanic Americans with type 2 diabetes. Journal of Periodontology, 79(4): 637–646.

Fang, J. -Y., Lin, C. -H., Huang, T. -H. and Chuang, S. -Y. (2019). *In vivo* rodent models of type 2 diabetes and their usefulness for evaluating flavonoid bioactivity. Nutrients, 11(3): 530. https://www.mdpi.com/2072-6643/11/3/530.

Gomes, A. C., Bueno, A. A., de Souza, R. G. M. and Mota, J. F. (2014). Gut microbiota, probiotics and diabetes. Nutrition Journal, 13(1): 1–13.

Grice, E. A. and Segre, J. A. (2012). The human microbiome: Our second genome. Annual Review of Genomics and Human Genetics, 13: 151–170.
Grice, E. A., Kong, H. H., Renaud, G., Young, A. C., Bouffard, G. G., Blakesley, R. W., Wolfsberg, T. G., Turner, M. L. and Segre, J. A. (2008). A diversity profile of the human skin microbiota. Genome Research, 18(7): 1043–1050.
Hur, K. Y. and Lee, M. -S. (2015). Gut microbiota and metabolic disorders. Diabetes & Metabolism Journal, 39(3): 198–203.
Jia, L., Li, D., Feng, N., Shamoon, M., Sun, Z., Ding, L., Zhang, H., Chen, W., Sun, J. and Chen, Y. Q. (2017). Anti-diabetic effects of Clostridium butyricum CGMCC0313. 1 through promoting the growth of gut butyrate-producing bacteria in type 2 diabetic mice. Scientific Reports, 7(1): 7046.
Jneid, J., Lavigne, J., La Scola, B. and Cassir, N. (2017). The diabetic foot microbiota: A review. Human Microbiome Journal, 5: 1–6.
Kahn, B. B. and Flier, J. S. (2000). Obesity and insulin resistance. J. Clin. Invest., 106(4): 473–481. https://doi.org/10.1172/jci10842.
Komaroff, A. L. (2017). The microbiome and risk for obesity and diabetes. Jama, 317(4): 355–356.
Lim, A. K. (2014). Diabetic nephropathy–complications and treatment. International Journal of Nephrology and Renovascular Disease, 361–381.
Löe, H. (1993). Periodontal disease: The sixth complication of diabetes mellitus. Diabetes Care, 16(1): 329–334.
Long, J., Cai, Q., Steinwandel, M., Hargreaves, M. K., Bordenstein, S. R., Blot, W. J., Zheng, W. and Shu, X. O. (2017). Association of oral microbiome with type 2 diabetes risk. Journal of Periodontal Research, 52(3): 636–643.
Mathers, C. D. and Loncar, D. (2006). Projections of global mortality and burden of disease from 2002 to 2030. PLoS Medicine, 3(11): e442.
Matsha, T. E., Prince, Y., Davids, S., Chikte, U., Erasmus, R. T., Kengne, A. and Davison, G. (2020). Oral microbiome signatures in diabetes mellitus and periodontal disease. Journal of Dental Research, 99(6): 658–665.
Nakatsuji, T., Chen, T. H., Two, A. M., Chun, K. A., Narala, S., Geha, R. S., Hata, T. R. and Gallo, R. L. (2016). Staphylococcus aureus exploits epidermal barrier defects in atopic dermatitis to trigger cytokine expression. Journal of Investigative Dermatology, 136(11): 2192–2200.
Negrato, C. A., Tarzia, O., Jovanovič, L. and Chinellato, L. E. M. (2013). Periodontal disease and diabetes mellitus. Journal of Applied Oral Science, 21: 1–12.
Negrini, T. d. C., Carlos, I. Z., Duque, C., Caiaffa, K. S. and Arthur, R. A. (2021). Interplay among the oral microbiome, oral cavity conditions, the host immune response, diabetes mellitus, and its associated-risk factors—An overview. Frontiers in Oral Health, 2: 697428.
Perry, R. J., Peng, L., Barry, N. A., Cline, G. W., Zhang, D., Cardone, R. L., Petersen, K. F., Kibbey, R. G., Goodman, A. L. and Shulman, G. I. (2016). Acetate mediates a microbiome–brain–β-cell axis to promote metabolic syndrome. Nature, 534(7606): 213–217.
Redel, H., Gao, Z., Li, H., Alekseyenko, A. V., Zhou, Y., Perez-Perez, G. I., Weinstock, G., Sodergren, E. and Blaser, M. J. (2013). Quantitation and composition of cutaneous microbiota in diabetic and nondiabetic men. The Journal of Infectious Diseases, 207(7): 1105–1114.
Salgaço, M. K., Oliveira, L. G. S., Costa, G. N., Bianchi, F. and Sivieri, K. (2019). Relationship between gut microbiota, probiotics, and type 2 diabetes mellitus. Applied Microbiology and Biotechnology, 103: 9229–9238.
Sardi, J. C., Duque, C., Camargo, G. A., Hofling, J. F. and Gonçalves, R. B. (2011). Periodontal conditions and prevalence of putative periodontopathogens and Candida spp. in insulin-

dependent type 2 diabetic and non-diabetic patients with chronic periodontitis—A pilot study. Archives of Oral Biology, 56(10): 1098–1105.

Schittek, B., Hipfel, R., Sauer, B., Bauer, J., Kalbacher, H., Stevanovic, S., Schirle, M., Schroeder, K., Blin, N. and Meier, F. (2001). Dermcidin: A novel human antibiotic peptide secreted by sweat glands. Nature Immunology, 2(12): 1133–1137.

Schommer, N. N. and Gallo, R. L. (2013). Structure and function of the human skin microbiome. Trends in Microbiology, 21(12): 660–668. https://doi.org/https://doi.org/10.1016/j.tim.2013.10.001.

Sedghi, L., DiMassa, V., Harrington, A., Lynch, S. V. and Kapila, Y. L. (2021). The oral microbiome: Role of key organisms and complex networks in oral health and disease. Periodontol 2000, 87(1): 107–131. https://doi.org/10.1111/prd.12393.

Shi, H., Kokoeva, M. V., Inouye, K., Tzameli, I., Yin, H. and Flier, J. S. (2006). TLR4 links innate immunity and fatty acid–induced insulin resistance. The Journal of Clinical Investigation, 116(11): 3015–3025.

Siljander, H., Honkanen, J. and Knip, M. (2019). Microbiome and type 1 diabetes. EBioMedicine, 46: 512–521.

Singer-Englar, T., Barlow, G. and Mathur, R. (2019). Obesity, diabetes, and the gut microbiome: An updated review. Expert Review of Gastroenterology & Hepatology, 13(1): 3–15. https://doi.org/10.1080/17474124.2019.1543023.

Stanley, S., Wynne, K. and Bloom, S. (2004). Gastrointestinal satiety signals III. Glucagon-like peptide 1, oxyntomodulin, peptide YY, and pancreatic polypeptide. American Journal of Physiology-Gastrointestinal and Liver Physiology, 286(5): G693–G697.

Swaney, M. H., Nelsen, A., Sandstrom, S. and Kalan, L. R. (2023). Sweat and Sebum preferences of the human skin microbiota. Microbiol. Spectr., 11(1): e0418022. https://doi.org/10.1128/spectrum.04180-22.

Tarr, J. M., Kaul, K., Chopra, M., Kohner, E. M. and Chibber, R. (2013). Pathophysiology of diabetic retinopathy. International Scholarly Research Notices, 2013.

Tisch, R. and McDevitt, H. (1996). Insulin-dependent diabetes mellitus. Cell, 85(3): 291–297.

WHO. (2023). Diabetes. World Health Organization. Retrieved 08 of April from https://www.who.int/health-topics/diabetes#tab=tab_1.

Wu, H., Tremaroli, V., Schmidt, C., Lundqvist, A., Olsson, L. M., Krämer, M., Gummesson, A., Perkins, R., Bergström, G. and Bäckhed, F. (2020). The gut microbiota in prediabetes and diabetes: a population-based cross-sectional study. Cell Metabolism, 32(3): 379–390. e373.

Zhang, S., Cai, Y., Meng, C., Ding, X., Huang, J., Luo, X., Cao, Y., Gao, F. and Zou, M. (2021). The role of the microbiome in diabetes mellitus. Diabetes Research and Clinical Practice, 172: 108645. https://doi.org/https://doi.org/10.1016/j.diabres.2020.108645.

CHAPTER 5

The Role of Microbiome in Non-Alcoholic Fatty Liver Disease (NAFLD)

Wesam Bahitham,[1,2] *Arwa Alghmdi,*[1] *Elana Hakeem,*[1] *Foad Sendi,*[1] *Abdullah Boubsit,*[1] *Eyad Alkhayat,*[1] *Ibrahim Omer,*[1] *Sharif Hala*[1] and *Alexandre Rosado*[2,*]

Introduction

Definition of Gut Microbiota (and its Function) and Dysbiosis

Gut microbiota (GM), or gut normal flora, is defined as the microorganisms, including bacteria, protozoa, fungi, archaea and viruses, that reside in the gastrointestinal tract of the human body. The gut microbiota consists of over 100 trillion bacteria along the gastrointestinal tract and has a 150-fold more complex genome than the human genome (Milosevic et al. 2019). GM plays a significant role in several physiological processes in the human body. For example, GM aids in the synthesis of vitamins, digestion, forming blood vessels (angiogenesis) and modulating the body's immune system. Moreover, GM has a significant role in creating short-chain fatty acids (SCFA), providing energy to the body. However, any disruption of the GM balance, i.e., dysbiosis, can disturb systemic pathological

[1] King Abdullah International Medical Research Center-WR, King Saud bin Abdulaziz University for Health Sciences, Ministry of National Guard for Health Affairs.
[2] Biological and Environmental Sciences and Engineering Division (BESE), King Abdullah University for Science and Technology (KAUST).
* Corresponding author: alexandre.rosado@kaust.edu.sa

processes, including the hepatic system through the gut-liver axis, the respiratory system, neurological system, endocrine system, and many other pathological diseases. Many factors contribute to the development of dysbiosis, such as host-related factors and, more importantly, environmental factors. Host-related factors include previous or current infections and inflammation, and genes. Environmental factors include high-fat and high-sugar diets, which are associated with the disruption of the barrier of the intestine, leading to more inflammation and eventually further damaging the functions of the gut, such as metabolism (Hrncir 2022).

Mechanism of Dysbiosis in NAFLD

When the gut microbiota is disrupted, several alterations in the body's physiological processes occur. First, there is an increase in insulin resistance, lipogenic genes, and reactive oxygen species (ROS), and a decrease in glutathione and beta-oxidation genes, leading to the disruption of physiological metabolism. Further damage can also affect the standard immunological mechanisms by increasing chemokines and the infiltration of immune cells, M1 macrophages and T-regulatory (Treg) cells, and decreasing mucosal immunity, intraepithelial lymphocytes (IELs) and cytotoxic T cells (Figure 1). Moreover, a significant alteration in the gut-derived metabolites occurs, including an increase or decrease in short-chain fatty acids and bile acids, an increase in trimethylamine N-oxide (TMAO), and a reduction in aryl hydrocarbon receptor (AhR), farnesoid X receptor (FXR) activation and tryptophan metabolites. These alterations allow further fat accumulation in the liver, consequently damaging normal hepatocytes. When hepatocytes are damaged, the liver can no longer adapt to the harmful fat accumulation, ultimately leading to lipotoxicity (Gupta et al. 2022).

Triggers of Dysbiosis

Multiple factors affect the typical composition of gut microbiota, including host-related and environmental factors, which are considered as the leading causes of dysbiosis. Host-related factors involve gastrointestinal or non-gastrointestinal infections and 'bactericidal' secretions from the gastric glands and liver. For example, gastric acid and bile influence gut microbiota; lysozymes, antibacterial lectins (Reg3γ) by paneth cells, or secretory immunoglobulin A (SIgA) by plasma cells can effectively result in dysbiosis (Hrncir 2022). Moreover, environmental factors, such as diet, are essential for developing dysbiosis in the host. Sugar-rich foods can

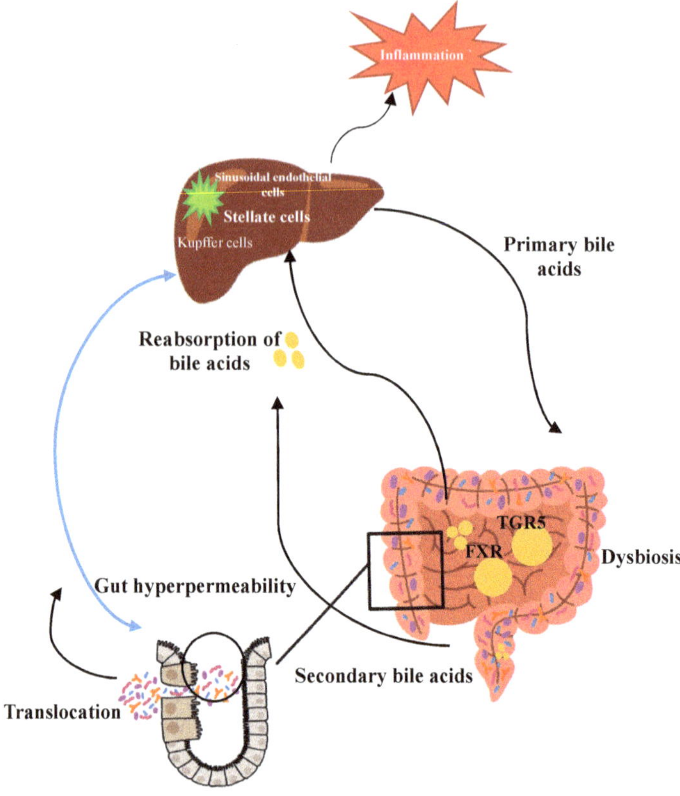

Figure 1. Implication of intestinal dysbiosis in NAFLD pathogenesis. Dysbiosis is associated with altered production of SCFA, altered bile acid metabolism, increased bacterial-derived ethanol, increased intestinal permeability, and upregulation of inflammatory factors. The consequences of dysbiosis affect normal liver physiology, given the gut and liver correlation. Hepatic lipogenesis and triacylglycerol increases, and lipid oxidation is decreased, leading to hepatic steatosis.

damage the intestinal wall barrier and cause an inflammation in the area. Furthermore, food additives, such as food preservatives, elicit increased growth of proteobacteria and affect the general health of the host. Other food additives include sweeteners and emulsifiers (in prepackaged and processed foods), which can cause intestinal inflammation and disturb the typical gut microbiota (Hrncir 2022).

Gut-Liver Axis

The gut and the liver are connected bidirectionally through a portal and biliary circulation. The superior mesenteric vein joins the splenic vein to form the portal vein, which is the central vein that drains blood to the liver. Furthermore, the liver is the main organ responsible for the synthesis of bile acid as it receives it from systemic circulation. Bile is then secreted into the gallbladder, through the common hepatic duct. It runs through the common bile duct, which runs from the liver to the duodenum. Thus, the liver and the gut are connected anatomically and functionally, forming the gut-liver axis (Milosevic et al. 2019).

Furthermore, bacterial translocation occurs whenever the gut permeability increases, enabling metabolites to reach the liver. When this event occurs, metabolites in the liver interfere with bile acid metabolism, leading to gut dysmotility, systemic inflammation and, eventually, gut dysbiosis. The gut-liver axis is responsible for several chronic hepatic diseases. For instance, in chronic hepatitis B (CHB), a decreased ratio of *Bifidobacteriacae/Enterobacteriaceae* is observed. Moreover, in non-alcoholic fatty liver disease (NAFLD) or non-alcoholic steatohepatitis (NASH), the *Firmicutes/Bacteriodetes* ratio is increased (Milosevic et al. 2019). Given the previous examples, scientists should have a better understanding of the impact of GM on the liver through the gut-liver axis; the latter should be taken into consideration while determining the pathogenesis of the disease (Milosevic et al. 2019).

Consequences of Dysbiosis

Toll-like receptors (TLR) and NOD-like receptors (NLR) modulate the host's immune system, so any disruption of the gut wall integrity activates TLR and NLR. Gut microbiota releases endogenous acetaldehyde, regardless of alcohol intake, which disrupts the gut wall integrity and imbalances the host's immune and metabolic systems (Hrncir 2022). TLR activation, cytokine release, and bile acid profile alterations result from increased ethanol levels. Increased levels of endogenous ethanol cause insulin resistance, lipid formation and intestinal hyperpermeability in patients diagnosed with NAFLD/NASH, aggravating exposure to pathogen-associated molecular patterns (PAMPs) and oxidative stress. Moreover, hepatic stellate cell activation by endotoxin/TLR4 signaling leads to fibrosis and cirrhosis, contributing to irreversible liver damage. Furthermore, gut microbiota plays a significant role in the modulation of bile acid conservation. Any disturbance in these processes is considered pathogenic (Milosevic et al. 2019).

Several animal studies have provided evidence of the role of the gut microbiome in the pathogenesis of NAFLD. Several potential links between the gut microbiome and NAFLD have emerged based on these animal studies. These mechanisms include dysregulation of methylamine metabolism, carbohydrate fermentation and generation of SCFAs, endogenous ethanol production, bile acid metabolism and amino acid metabolism. Moreover, intestinal barrier dysfunction and lipopolysaccharide-induced activation of toll-like receptor pathways may contribute to NAFLD (Sharpton et al. 2019).

Methylamine Metabolism

Metabolism of dietary methylamines, specifically choline, by the gut microbiome may contribute to the development of NAFLD. A choline deficiency can contribute to the development of fatty liver disease through multiple mechanisms, including abnormal phospholipid synthesis, defective very low-density lipoprotein secretion, and alterations in the enterohepatic circulation of bile acids (Sherriff et al. 2016). Moreover, in mice consuming a high-fat diet, choline increases gut microbial metabolism, and subsequent development of hepatic steatosis is noticed (Dumas et al. 2006). However, additional studies are required to determine the role of the gut microbial metabolism of choline in the pathogenesis of NAFLD.

Bile Acid Transformations by Gut Microbiota

Bile acids are endocrine molecules that facilitate the absorption of fat-soluble nutrients, while regulating numerous metabolic processes, including glucose, lipid and energy homeostasis. Microbes in the gut are responsible for the intestinal modification of bile acids and the regulation of several hepatic critical enzymes involved in bile acid synthesis. Alterations in bile acid homeostasis are often associated with metabolic diseases (Molinaro et al. 2018). Additionally, bile acids play an essential role in shaping the gut microbiome membership, which is a result of these bile acids contributing to the prevention of intestinal bacterial growth, both directly through membrane-damaging effects and indirectly through the induction of antimicrobial protein expression (Mouzaki et al. 2016, Slijepcevic and van de Graaf 2017). Moreover, studies on humans have noted elevated levels of total serum bile acids in adults with NAFLD (Ferslew et al. 2015, Mouzaki et al. 2016, Jiao et al. 2018).

Endogenous Ethanol Production

For individuals who do not regularly consume alcoholic beverages and those who do not have access to alcohol, circulating alcohol comes from the diet and the intestinal bacteria. For instance, after consuming fruits, methanol concentration in the human body increases (Lindinger et al. 1997). The pathomechanisms of alcohol in liver diseases have been intensively studied in alcoholic liver disease. It has been long observed that ethanol stimulates hepatic fatty acid synthesis (Lieber and Schmid 1961). The oxidation of alcohol to aldehyde by ADH concomitantly reduces NAD to NADH. The increased NADH levels promote fatty acid synthesis and oppose lipid catabolism, leading to hepatocyte fat accumulation (Lieber and Schmid 1961, Galli et al. 1999). In a recent study, elevated ADH levels strengthened the connection between alcohol metabolism and steatosis (Baker et al. 2010). Furthermore, the gut microbiome can generate ethanol from dietary precursors, and manipulating the gut microbiome is known to alter endogenous ethanol production (Elshaghabee et al. 2016).

Amino Acid Biosynthesis and Metabolism

Amino acid homeostasis is influenced by the gut microbiome, in part due to the biosynthesis and metabolism of aromatic amino acids (AAAs) and branched-chain amino acids (BCAAs) (Sharpton et al. 2019). Several cohort studies have identified elevated serum BCAA levels as a potential biomarker for insulin resistance (Menni et al. 2013). In a cohort study by Hoyles and collaborators, women with NAFLD had significant alterations in the gut metagenome, including differences in BCAA and AAA pathways and serum metabolome (Hoyles et al. 2018). This study proves how integrated analyses in human subjects can facilitate the identification of microbial-driven mechanistic pathways in NAFLD.

Gut Microbiota and its Metabolites (Microbial, Dietary and Host-derived)

Short Chain Fatty Acids (SCFAs)

Short-chain fatty acids (SCFAs) are considered to be an essential class of microbial metabolites that impact the maintenance of health and the development of diseases. When consuming food, indigestible complex polysaccharides travel through the gastrointestinal tract, reaching the intestine where the intestinal microbiota reside. SCFAs are mainly produced by the intestinal microbiota, through the anaerobic fermentation of indigestible dietary polysaccharides (Agus et al. 2021, Visekruna and Luu 2021). Butyrate, acetate and propionate are examples of SCFAs, which have many cellular effects, depending on the type of cells they are linked to. They have been shown to have an agonist effect on eukaryotic G-protein-coupled receptors (GPRs), which play an essential role in cellular signaling pathways.

An example is acetate and propionate binding to GPR41 and GPR43 expressed on colonocytes, which activates the mitogen-activated protein kinases (MAPKs) pathway, leading to an inflammatory response (Visekruna and Luu 2021). Furthermore, SCFAs binding to GPR109a expressed on dendritic cells in the intestine-induced regulatory T-cells proliferation and activation. Moreover, their small size allows SCFAs to enter the cells directly by simple diffusion. They can be carried by sodium-coupled transporters and enter the cell indirectly, where they can act on the cell organelles.

Short-chain fatty acids, like butyrate, can be a therapeutic agent in certain diseases. For instance, in the intestine, butyrate enhances the healing process of damaged intestinal cells, promotes barrier function and maintains intestinal homeostasis. In addition, a study conducted on hyperglycemic mice showed that butyrate could promote the intestinal expression of glucagon-like peptide 1 (GLP-1). GLP-1, a primary incretin hormone, is predominant in maintaining glucose homeostasis. It also stimulates insulin release, inhibits glucagon release and gastric emptying, and suppresses the appetite, ultimately decreasing glucose levels in the blood.

Moreover, butyrate can also ameliorate atherosclerosis by affecting cholesterol and phospholipid homeostasis. It up-regulates ABCA-1 gene expression, which encodes cholesterol efflux regulatory protein (CERP). This protein promotes cholesterol efflux, thus alleviating atherosclerosis (Ma et al. 2022).

There is increasing evidence suggesting that Short-chain fatty acids (SCFAs) impact the physiological functions of the liver, either directly or indirectly. Glucagon-like peptide 1 receptor (GLP-1r) is located on hepatocytes and is a critical factor in regulating hepatic lipid metabolism. GLP-1r expression decreases in people with non-alcoholic fatty liver disease (NAFLD). SCFAs can promote GLP-1 secretion, which activates GLP-1r on hepatocytes; thus, supplementing with SCFAs, such as acetate and propionate, can provide potential benefits to NAFLD patients. On the contrary, SCFAs can directly affect the liver, promoting NAFLD. The mechanism through which SCFAs cause NAFLD is increasing triglyceride accumulation and hepatic gluconeogenesis, caused by acetate and propionate respectively (Ding et al. 2019).

Ethanol and its Metabolites

Alcohol metabolism is a natural process in the liver and generates products that damage the liver (Cederbaum 2012). While some ingested alcohol might be absorbed in the mouth, most of it is absorbed and metabolized in the gastrointestinal tract, especially the stomach and the small intestine. Following absorption, about 10% of the absorbed alcohol is excreted unchanged from the body, through breath, sweat and urine (Hyun et al. 2021); this leaves the body with about 90% of the ingested alcohol that circulates throughout the body, which is eventually delivered to the hepatic portal vein. The liver is a crucial player in alcohol metabolism as it possesses high levels of enzymes that can metabolize alcohol (Cederbaum 2012). Alcohol can undergo metabolism inside the liver by one of the two interconnected pathways – oxidative and non-oxidative (Cederbaum 2012).

The oxidative pathway is the main pathway for alcohol metabolism, consisting of two steps. First, the reversible oxidation of alcohol into acetaldehyde is performed by the main oxidizing enzyme in the liver, alcohol dehydrogenase (ADH) (Lieber 1997). It can be found in the cytosol of hepatocytes and has a high affinity for alcohol. Overconsumption of alcohol is linked to the activation of cytochrome P450 2E1 (CYP2E1) instead of ADH. CYP2E1 is found in peroxisomes (Cederbaum 2012). Similar to ADH, active CYP2E1 contributes to the formation of acetaldehyde. However, it leads to the formation of reactive oxygen species (ROS), such as hydroxyethyl, superoxide anion and hydroxyl radicals, all of which affect the liver negatively (Leung and Nieto 2013).

Furthermore, this oxidation step involves an intermediate electron carrier, nicotinamide adenine dinucleotide (NAD+), which is reduced by two electrons to form NADH. This results in a highly reduced cytosolic environment in hepatocytes, which exposes the liver to damage from the products of alcohol metabolism (Harjumaki et al. 2021). Another enzyme that can oxidize alcohol to acetaldehyde is catalase, which is also found in peroxisomes. This minor pathway plays a prominent role in fasting (Harjumaki et al. 2021). The second step of the oxidative pathway is acetaldehyde oxidation, forming acetate and NADH; this is done by the aldehyde dehydrogenase (ALDH) enzyme found in the mitochondria (Cederbaum 2012).

Moreover, acetaldehyde can be metabolized by the enzyme CYP2E1. Most of the produced acetate then travels to extrahepatic tissues, such as the heart and the brain, where it is metabolized into carbon dioxide (CO_2). It can also be metabolized into fatty acids and water in extrahepatic tissues, and not in the liver (Rehm et al. 2010).

The second mechanism by which alcohol is metabolized is through non-oxidative pathways. Unlike the oxidative pathway, only minor portions of alcohol undergo metabolism through these pathways (Hyun et al. 2021). Nonetheless, they are essential because their products can be used in pathological and diagnostic scenarios. One of these pathways leads to the formation of phosphatidyl ethanol, which results from the formation of fat molecules containing phosphorus. This pathway is vital in cellular communication. The enzyme phospholipase D (PLD) is essential for this step. PLD breaks down phospholipids to produce phosphatidic acid (PA). Reactions that involve PLD usually occur in the presence of a high concentration of alcohol. Higher phosphatidyl ethanol levels can be detected after ingesting large amounts of alcohol as it is metabolized poorly. Another non-oxidative pathway is the formation of fatty acid ethyl esters (FAEEs). FAEEs can be detected in serum following alcohol intake. The role of FAEEs in alcohol-induced liver disease still remains unclear (Zakhari 2006).

The liver is the organ affected by alcohol the most, for it is the primary site for alcohol metabolism. Products of alcohol metabolism are highly associated with liver damage and alcoholic liver disease (ALD) (Szabo 2015). Following chronic alcohol consumption, the first sign of liver damage is the development of fatty liver, which is the accumulation of fats around the liver. Fatty liver has been observed in about 90% of alcoholics (Wilfred de Alwis and Day 2007). The development of fatty liver is usually followed by various mechanisms that lead to liver damage. Inflammation, fibrosis and cirrhosis are some mechanisms through which the liver is affected. Shifting in the redox state of hepatocytes plays a significant role in the development of a fatty liver. Instead of oxidizing alcohol, this shift results in a fat-storing liver, which can be observed as the accumulation of lipids in hepatocytes (Zakhari 2006). Another mechanism by which fatty liver is developed is the activation of proteins that aid in fatty acid synthesis (Nagy 2004).

Another consequence of excessive alcohol consumption is alcoholic hepatitis, which is characterized by inflammation and necrosis (Lucey et al. 2009). Alcohol disrupts the intestinal barrier. This increases permeability and exposure to gut bacteria and lipopolysaccharides (LPS). LPS actives Kupffer cells – resident macrophages that play a vital role in maintaining liver function (Nguyen-Lefebvre and Horuzsko 2015). Activated Kupffer cells release inflammatory cytokines and ROS excessively. If persistent, this irregular inflammatory response results in liver damage and tissue necrosis. Furthermore, fatty liver also plays a role in the inflammatory processes of the liver (Zhu et al. 2012). Glucocorticoids are used in moderate to severe alcoholic steatohepatitis (ASH), which supports the crucial role of inflammation in ALD (Frazier et al. 2011). Furthermore, elevated oxidative stress has been observed in alcoholic hepatitis, which powerfully demonstrates the role of ROS in the development of alcoholic hepatitis (Meagher et al. 1999).

Mechanisms described earlier, such as the shift of oxidative stress, inflammation, and accumulation of fats in hepatocytes, are all linked to the pathogenesis of alcoholic cirrhosis and, eventually, hepatocellular carcinoma (Hyun et al. 2021).

Amino Acid Metabolism and its Byproducts

Twenty amino acids are classified as essential and non-essential, depending upon their source. The nine essential amino acids (lysine, histidine, methionine, phenylalanine, tryptophan, threonine, valine, leucine and isoleucine) are not synthesized or are minimally synthesized by the human body, and are acquired from the diet. In contrast, the eleven non-essential amino acids (glycine, arginine, cysteine, tyrosine, serine, alanine, proline, asparagine, aspartate, glutamine and glutamate) are normally synthesized by the human body. These nitrogenous substances are vital nutrients to support the growth of gut bacteria and their host, and to regulate energy and protein homeostasis in the human body. Generally, amino acids and peptides are products of endogenous and dietary proteins hydrolyzation, which are mediated by different types of host- and bacteria-derived peptidases and proteases (Dai et al. 2011). The gut bacteria and the host digestive system further utilize these products. The interplay between gut microbes and the host metabolism is complex, since the bacteria can synthesize and use amino acids to produce metabolites that affect the host's amino acid intake (Portune 2016). While relatively high amounts of amino acid concentrations are present in the small intestine, the greater concentration of microorganisms is in the large intestine, accompanied by high rates of protein fermentation. Therefore, enterocytes in the small intestine can degrade many non-essential amino acids, with limited oxidation of essential amino acids.

On the other hand, the essential amino acids are not significantly absorbed by the colonic mucosa. Instead, they are metabolized by the colonic microbiota, which can metabolize all kinds of amino acids (Dai et al. 2011, Neis et al. 2015). For the initial step of catabolism, a microbe can use either deamination or decarboxylation to produce carboxylic acid and ammonia or amine and carbon dioxide respectively (Oliphant and Allen-Vercoe 2019). Deamination is considered the more common pathway since it produces more short-chain fatty acids. The product of this pathway, ammonia, can negatively affect the host as it can decrease short-chain fatty acid degradation by the intestinal epithelial cells and inhibit mitochondrial oxygen consumption. However, it is highly regulated by the host and the gut microbiota. The intestinal epithelial cells control ammonia concentration by transforming it into citrulline and glutamine or releasing it into the bloodstream (Oliphant and Allen-Vercoe 2019). A study suggests that citrulline can prevent liver fat accumulation

(Jegatheesan et al. 2016). Glutamine, the most abundant free amino acid in the human body, is vital for regulating systemic pH and the urea cycle in the liver (Haussinger and Schliess 2007). The next step is bacterial protein fermentation. This reaction happens at higher rates in the large intestine due to the high pH levels and the low concentration of carbohydrates. Colonic microorganisms favor certain amino acids, such as lysine, glycine and arginine, as well as branched-chain amino acids (valine, leucine and isoleucine). Generally, this reaction generates ammonia, short-chain fatty acids and branched-chain fatty acids as the end products (Hrncir 2022).

Essential Amino Acids

Lysine, a primary amino acid, is catabolized by lysine decarboxylase to produce acetate, butyrate and cadaverine (Oliphant and Allen-Vercoe 2019). Acetate can enhance lipid output and fatty acid oxidation, inhibiting synthesis. Thus, it reduces hepatic triglyceride content (Liu et al. 2019). Moreover, butyrate is protective against fatty liver and insulin resistance in diet-induced obese mice (Mollica et al. 2017). Little is known about cadaverine, but it can be toxic in high concentrations, as it intensifies histamine toxicity (Oliphant and Allen-Vercoe 2019); it also assists *Escherichia coli* with an acid resistance mechanism (Portune et al. 2016). Another amino acid, arginine, is an essential amino acid catabolized into agmatine, putrescine, spermidine and spermine, via arginine decarboxylase (Oliphant and Allen-Vercoe 2019). Many species of bacteria are involved in this reaction, such as *Escherichia coli, Klebsiella pneumoniae* and *Selenomonas ruminantium* (Dai et al. 2011). Agmatine affects other polyamines by reducing their synthesis and promoting their degradation. This can have positive or negative effects, depending upon the situation. For example, agmatine can improve metabolic syndrome by decreasing fatty acid metabolism and reducing body weight. On the other hand, it can prove to be harmful by decreasing putrescine, spermine and spermidine, since these three polyamines increase intestinal mucus secretion and improve gut integrity (Oliphant and Allen-Vercoe 2019). Histidine decarboxylase catalyzes the production of histamine, another amino acid, from histidine (Portune et al. 2016). The main bacterial species involved in this reaction is *Clostridium perfringens* (Dai et al. 2011). Bacterially produced histamine is associated with the inhibition of cytokines and interleukins, as well as preventing the translocation of gut bacteria. Further, it is a known neurotransmitter for pain and memory (Oliphant and Allen-Vercoe 2019). Other bacterial species, such as *Fusobacterium nucleatum, Fusobacterium varium, Klebsiella pneumoniae* and *Acidaminococcus fermentans*, produce urocanate and ammonia

(Dai et al. 2011). Methionine, a sulfur-containing amino acid, is catabolized to methanethiol and, to a lesser extent, to propionate and butyrate. Bacterial species, such as *Proteobacteria phylum, Klebsiella pneumoniae, Clostridium* species and *Bifidobacterium* species, have the required enzymes to make this reaction happen (Dai et al. 2011, Oliphant and Allen-Vercoe 2019). Methanethiol, also known as methyl mercaptan, is one of the chemicals responsible for the fusty odor of flatus. An old study suggests that blood methanethiol levels can be used to evaluate patients with possible hepatic encephalopathy (McClain et al. 1980).

Furthermore, a study showed that producing propionate from nutrient fermentation can dampen malignant cell proliferation in the liver tissues (Bindels et al. 2012).

Phenylalanine is an aromatic amino acid catabolized into phenylethylamine and trans-cinnamic acid, also known as phenyl acrylic acid, via some *Peptostreptococcus* species (Dai et al. 2011, Oliphant and Allen-Vercoe 2019). Little is known about the effects of these products, but phenylethylamine and phenethylamine enhance the release of catecholamine and serotonin to elevate mood and increase attention (Shimazu and Miklya 2004). A recent experiment suggests that oral phenyl acrylic acid derivative is an excellent anti-hepatic fibrosis candidate (Xue et al. 2022). Another aromatic amino acid is tryptophan. Tryptophan metabolism is a rare action in the gut microbiota, yet *Clostridium sporogenes* converts decarboxylate tryptophan into tryptamine and indole (Dai et al. 2011, Oliphant and Allen-Vercoe 2019). Tryptamine is a neurotransmitter that induces serotonin release to modify gastrointestinal motility and potentially plays a role in inflammatory bowel disease pathology (Portune et al. 2016). Next, threonine, a hydroxylic amino acid, is catabolized by many bacterial species to acetate, propionate and butyrate (Dai et al. 2011, Oliphant and Allen-Vercoe 2019); *Peptostreptococcus* and *Clostridium* species are involved in this metabolism (Dai et al. 2011).

Along with arginine and serine, *Clostridium sticklandii* prefers to use threonine as a carbon and energy source (Portune et al. 2016). The final three essential amino acids – leucine, isoleucine and valine – are branched-chain amino acids. *Clostridium* species are mainly responsible for the catabolism of these amino acids (Dai et al. 2011). Their catabolism produces branched-chain fatty acids – isovalerate, 2-Methylbutyrate and isobutyrate respectively (Fan et al. 2015). Branched-chain fatty acids can be used as a biomarker for protein levels, given their ability to regulate lipid and glucose metabolism in hepatocytes (Oliphant and Allen-Vercoe 2019). A study states that a leucine-restricted diet improves insulin sensitivity (Xiao et al. 2011). An increase in aromatic and branched-chain amino acids is linked to the development of insulin resistance and type 2 diabetes mellitus (Wang et al. 2011).

Non-essential Amino Acids

The first amino acid is glycine. This non-essential amino acid is transported to the intestine by glycine-preferring transporters throughout the borders of the intestinal cell membrane in a gradient-dependent manner (Fan et al. 2015). Glycine can be fermented by *Fusobacterium nucleatum, Escherichia coli, Clostridium sticklandii, Clostridium difficile* or *Clostridium perfringens* to produce acetate and methylamine (Dai et al. 2011, Oliphant and Allen-Vercoe 2019). Methylamine is converted into formaldehyde through methylamine dehydrogenase. The next amino acid is arginine. It is an essential amino acid catabolized to agmatine, putrescine, spermidine and spermine, via arginine decarboxylase (Oliphant and Allen-Vercoe 2019). Many species of bacteria are involved in this reaction, such as *Escherichia coli, Klebsiella pneumoniae,* and *Selenomonas ruminantium* (Dai et al. 2011). Agmatine affects other polyamines by reducing their synthesis and promoting their degradation. This can have positive or negative effects, depending upon the situation. For example, agmatine can improve metabolic syndrome by decreasing fatty acid metabolism and decreasing body weight. On the other hand, it can be harmful by decreasing putrescine, spermine and spermidine since these three polyamines increase intestinal mucus secretion and improve gut integrity (Oliphant and Allen-Vercoe 2019). Next, cysteine is a sulfur-containing amino acid that produces acetate, butyrate and hydrogen sulfide by the action of cysteine desulfhydrase enzyme (Portune et al. 2016). Excessive concentrations of hydrogen sulfide harms the gut. It acts on the disulfide bonds to enhance mucin degradation and inhibit mitochondrial cytochrome C oxidase (Oliphant and Allen-Vercoe 2019). Moreover, the liver is the main regulator of endogenous and exogenous hydrogen disulfide metabolism. The synthesis of lipoproteins, insulin sensitivity and glucose metabolism are all impacted by hepatic H2S metabolism. Thus, numerous liver conditions, including hepatic cirrhosis and hepatic fibrosis, may be impacted by abnormalities in hepatic H2S metabolism (Begley et al. 2005). Tyrosine, a non-essential aromatic amino acid, is catabolized to produce tyramine, phenols and p-coumarate, by tyrosine decarboxylase tyrosine phenol-lyase (Portune et al. 2016, Oliphant and Allen-Vercoe 2019). Tyrosine is a precursor of L-DOPA – a precursor for dopamine, epinephrine and norepinephrine (Portune et al. 2016). Tyrosine is a neurotransmitter that facilitates the release of norepinephrine and serotonin, leading to increased glucose levels and constriction of peripheral vessels. It is also associated with cheese reaction hypertensive crisis in allergic patients, when produced by *Enterococcus* and *Enterobacteriaceae* through decarboxylation (Oliphant and Allen-Vercoe 2019). Phenols and p-coumarate (phenolic metabolites) are produced by *Enterobacteriaceae* and some *Clostridium* clusters. Furthermore, they decrease the gut epithelium's integrity, inhibit intestinal epithelial cell proliferation and damage renal tubular cells. Serine is a hydroxylic amino acid metabolized by different types of gut bacteria, such as *Fusobacterium varium*, to produce acetate, butyrate and lactate (Dai et al. 2011). Next, alanine is primarily metabolized by different *Clostridium species* to produce butyrate (Dai et al. 2011, Oliphant and Allen-Vercoe 2019). *Campylobacter jejuni* metabolizes proline to produce acetate or glutamate. Moreover, proline dehydrogenase inhibition leads to the inhibition of urease and ATP synthesis from proline and eventually causes a growth-inhibitory effect mediated by *Helicobacter pylori*. The next amino acid is asparagine. *Fusobacterium nucleatum, Escherichia coli, Klebsiella pneumoniae, Campylobacter jejuni* and *Clostridium perfringens* convert asparagine to aspartate and produce ammonia (Dai et al. 2011). Aspartate is metabolized by different species of bacteria, such as *Campylobacter jejuni, Acidaminococcus fermentans* and *Bacteroides fragilis*, to produce alanine and organic acids (fumarate and oxaloacetate). Aspartate is a key factor in *Campylobacter jejuni* survival during nutrient limitation, since it controls the utilization of amino acids and ATP production (Dai et al. 2011). The last amino acid is glutamate. It is the ionic form of glutamic acid. Glutamate decarboxylase is the enzyme that catalyzes the production of gamma-aminobutyric acid (GABA), a known inhibitory neurotransmitter produced microbiologically by *Escherichia coli* and *Clostridium perfringens* to maintain intracellular pH homeostasis through proton consumption. Afterward, GABA is either exported from the cell via a secondary active transporter or remains in the cell, to be metabolized via the GABA shunt pathway to produce succinate (Portune et al. 2016). GABA depletion is linked to depression and anxiety, but has immunomodulatory properties (Oliphant and Allen-Vercoe 2019). Furthermore, glutathione consists of cysteine, glycine and glutamic acid in most cells with the same concentration as potassium, glucose and cholesterol since it requires high levels of metabolic activities. Moreover, it protects cellular molecules from endogenous and exogenous reactive oxygen and nitrogen, detoxifies xenobiotic compounds, enhances the transport of toxins through the plasma membrane, and neutralizes the free radicals produced by liver metabolism (Pizzorno 2014). In *L. monocytogenes*, the tripeptide functions as a signaling molecule to trigger the expression of the virulence thermoregulatory (Gupta et al. 2022). Overall, the control of nitrogen and energy balance in tissues and throughout the body, including the flow of the citric acid cycle and the urea cycle, as well as protein synthesis and breakdown, relies greatly on glutamine/glutamate, aspartate and glutathione (Dai et al. 2011). In conclusion, nitrogen and amino acid metabolism in the gut microbiota and the underlying metabolic interactions with the host have both negative and positive effects on the microbiome and the host's energy balance, protein regulation, and overall function.

Bile Acid as Gut Messenger

Bile acids are produced from cholesterol, and are classified as primary and secondary, depending upon their origin. Cholic acid and chenodeoxycholic acid are the primary bile acids synthesized in the liver and stored in the gallbladder. Before leaving the liver, they are conjugated to glycine or taurine to form bile salts, which increases their solubility, minimizes passive absorption, and resists cleavage by pancreatic peptides. Furthermore, microorganisms in the gut, especially in the colon, can remove the conjugated molecules to regenerate bile acids through an alternative route. Secondary bile acids are formed by removing a hydroxyl group from primary bile acids. Thus, cholic acid and chenodeoxycholic acid are converted into deoxycholic acid and lithocholic acid respectively. *Blautia, Rumminococcaceae,* and members of *Clostridium* clusters, including *C. scindens, C. hiranonis, C. hylemonae* and *C. sordelli,* are the most common bacterial species which produce secondary bile acids (Ridlon et al. 2014). Generally, bile acids facilitate absorption and digestion, regulate hepatic lipids and glucose, maintain metabolic homeostasis, and prevent the accumulation of cholesterol and toxic metabolites (Chiang 2013). For a better understanding of how bile acids act, it is critical to understand the enterohepatic circulation of bile acids and their metabolism. After a meal, the intestine produces cholecystokinin to stimulate the contraction of the gallbladder for the secretion of bile acids in the digestive system. The unconjugated bile acids are passively reabsorbed in the upper intestine, while the conjugated bile acids are reabsorbed in the terminal ileum. Then, the bile acids aid in the intestinal absorption of lipids and lipophilic vitamins, and enhance the emulsification of dietary fats. Following this, the bacteria in the distal intestine deconjugate and dehydroxylate bile acids to produce secondary bile acids. Most of the lithocholic acid is excreted through feces, whereas deoxycholic acid is reabsorbed in the colon and recycled to the liver, along with the primary bile acids. Moreover, bile acids are signaling molecules that activate three nuclear receptors and G protein-coupled receptors. Activation of these receptors starts the signaling cascade and activates expression genes to regulate energy consumption and metabolism of lipids and carbohydrates (Ramirez-Perez et al. 2017). The three nuclear receptors are farnesoid X receptor (FXR), pregnane X receptor (PXR) and vitamin D receptor (VDR). FXR affects bile acids' synthesis, biliary secretion, intestinal absorption, hepatocyte uptake, and detoxification. Additionally, it regulates lipid and glucose hepatic metabolism. FXR also plays an important role in preventing cholestasis (Chiang 2013). The next nuclear receptor is PXR. The PXR-dull mice experiment has shown that PXR is activated by lithocholic acid and its metabolites. Therefore, the receptor regulates the genes involved in lithocholic acid synthesis, transport and metabolism (Staudinger et al. 2001). PXR plays a significant role in detoxifying bile acids, drugs and toxins since it is extensively expressed in the liver and the gut (Chiang 2013). Looking ahead, the hepatotoxic and enteric carcinogenic compound lithocholic acid promotes its own metabolism by interacting with the vitamin D receptor (VDR), with lithocholic acid acetate being the most powerful activator of VDR (Adachi et al. 2005). The effect is mediated through CYP3A expression. When lithocholic acid and vitamin D bind to VDR, a feed-forward catabolic cascade is activated to increase the expression of CYP3A. Subsequently, lithocholic acid is detoxified (Makishima et al. 2002). To summarize, the activation of FXR, PXR and VDR regulates bile acids, lipids, glucose, toxins and energy metabolism. The other receptor is the G protein bile acid receptor-5 (TGR5). TGR5 is located in the apical membrane of the intestine. The activation of TRG5 stimulates gallbladder refilling and GLP-1 secretion (Chiang 2013). The secretion of GLP-1 improves insulin sensitivity. Moreover, the activated TRG5 stimulates energy metabolism and protects the liver and the intestine against inflammatory diseases. This happens by inhibiting macrophages from producing proinflammatory cytokine and inhibiting atherosclerosis (Thomas et al. 2008). Bile acids have direct as well as indirect antimicrobial effects that inhibit bacterial overgrowth. The indirect effects result from the activation of FXR and TGR-5 to regulate the overgrowth of intestinal microbiota, thus protecting the liver and the intestine from inflammation (Torres-Fuentes et al. 2017). Deoxycholic acid is the most potent antimicrobial agent among all amino acids because of its hydrophobicity and the ability to destroy the bacterial membrane (Begley et al. 2005). Decreased levels of bile acids in the gut enhance the concentration of gram-negative bacteria, leading to increased lipopolysaccharides production and possible pathogens (Ridlon et al. 2014). In conclusion, bile acids activate nuclear and G-protein coupled receptors to facilitate absorption, regulate lipid and glucose metabolism, improve insulin sensitivity, and maintain energy homeostasis. Also, bile acids protect the body by detoxifying drugs and toxins, inhibiting inflammatory cascades, and preventing bacterial overgrowth in the intestine.

Therapeutic Approaches

Gut Microbiome Targeted Therapy in the Management of NAFLD

Over the past decade, many clinical trials have been conducted to optimize and translate the beneficial effects of probiotics in diseased human conditions. However, translating microbiota-based therapy is still under development. Furthermore, the influence of gut microbiota has led to numerous preclinical and clinical studies for the effective prevention and treatment of NAFLD, NASH and NAFLD-HCC (Gupta et al. 2019). Nevertheless, various probiotics, such as *Lactobacilus*, *Bifidobacterium* and *Pediococcus*, have demonstrated beneficial effects in abrogating NAFLD in preclinical models. The administration of probiotics in rodents restores microbial homeostasis in the gut, reducing lipogenesis, subsequently lowering inflammation in the liver and eventually abolishing NAFLD (Lee et al. 2021, Machado et al. 2021, Yu et al. 2021). In a recent systematic review and meta-data analysis, Sharpton et al. (2019) found that probiotics or synbiotics are positively associated with improved liver function enzymes, hepatic steatosis, and liver stiffness measurements (Sharpton et al. 2019). On another note, recent evidence suggests that transplanting microbiota based on the unique gut environment offers prospects for personalized microbiome reconstitution (Maldonado-Gomez et al. 2016). However, due to their relatively small scale and heterogenous population, microbiota-based targeted therapy still needs to be expanded to preliminary clinical studies, in addition to random dosing and diverse endpoints. Other strategies focused on altering gut microbial compositions for beneficial effects in NAFLD are prebiotics, prebiotic and probiotic combinations (synbiotics), antibiotics, and fecal matter transplantation (FMT). Finally, due to the rapid increase in the prevalence and incidence of NAFLD, there is a need to develop new drugs for treatment. However, considering the complexity of the gut ecosystem and the substantial variation between rodent models and human disease conditions, translating the diseased rodent models of gut microbiome studies into human disease conditions is crucial.

Antibiotics, Prebiotics, Probiotics and Synbiotics

Solitharomycin, a promising macrolide antibiotic, has recently shown marked reduction in non-alcoholic fatty liver disease (NAFLD) Activity Score (NAS) and ALT [NCT02510599], providing evidence for the close inter-relationship between gut microbiota and the pathogenesis of NAFLD. The Food and Agricultural Organization of the United Nations and the World Health Organization define probiotics as "live microorganisms which, when administered in adequate amounts, confer a health benefit on the host" (2001). Probiotics are another potential therapy for NAFLD, mainly because they use beneficial gut microbes to counteract those involved in the progression of the disease. Several randomized controlled trials (RCTs) have assessed probiotics, with multiple reported positive outcomes. One RCT, for instance, has reported decreased AST levels and intra-hepatic triglycerides in patients with NASH (Wong et al. 2013).

Another RCT has reported similar outcomes in patients with NAFLD (Ahn et al. 2019), providing evidence of the efficacy of probiotics in the early stages of the disease. Furthermore, a third study reported similar clinical findings amongst children with NASH, further elucidating that probiotics correlate positively, regardless of age or stage of disease (Alisi et al. 2014). However, one particular concern with probiotics is the inconsistency of probiotics used, necessitating more extensive clinical trials with well-defined clinical outcomes that may approve clinical reasoning. Next-generation probiotics are promising candidates for treating NAFLD, especially considering their disease specificity.

While probiotics appear to have a significant impact on the progression of NAFLD, prebiotics, best defined as "substrate[s] that [are] selectively utilized by host microorganisms conferring a health benefit" (Gibson et al. 2017), appear to have a positive yet limited impact. As such, one RCT reported AST reductions but no plasma lipid reductions in a small cohort of biopsy-confirmed NASH patients (Daubioul et al. 2005). *Bifidobacterium* gut microflora appears to be the most beneficial, with a larger RCT suggesting reduced intra-hepatic fat accumulation in patients with NASH (Bomhof et al. 2019). Despite recent advances, prebiotics still need to be improved through research, warranting further and more extensive studies to be conducted.

Microbiota-based Diagnostics and Biomarkers

The diagnosis and follow-up of patients with NAFLD require liver biopsy and imaging. A biopsy is a crucial procedure as it enables the classification of tissue samples based on histology, which can provide additional clinical information that may influence a patient's prognosis or likelihood of remission. However, imaging and biopsy remain expensive procedures that require time and effort, limiting their upscale to community-based screening and extensive patient cohort follow-up. As discussed earlier, NAFLD appears to be associated with signature microbiota. Considering the large microbiota datasets generated from patients with NAFLD, such correlates cannot be based on a single organism but rather on large data machine learning models that include thousands of data points. One such example is a recently developed microbiota-based model that accurately predicted NAFLD progression and liver accumulation in a 4-year follow-up, showing evidence for the future use of such models to predict NAFLD early (Leung et al. 2022).

Microbiota-based Interventions

Bovine colostrum (BC) is best defined as the nutrient-rich, immunologically active bovine milk secreted within the first 72 hours after birth. Its immunological, besides nutritional, role is particularly evident as it provides pre-formed immunoglobulins (Ghosh and Iacucci 2021), specifically IgG, that noticeably reduce the translocation of lipopolysaccharides (LPS) from the gut flora. The decrease in inflammation is believed to offer significant protection, making ongoing clinical studies on IMM124-E particularly promising (Immuron Ltd, Melbourne, Australia). IMM124-E is an oral, colostrum-derived IgG agent that has shown considerable improvement in the GLP-1 level, adiponectin level, and Tregs count in phase I/II clinical trials on patients with NASH (Mizrahi et al. 2012). Its effect on GLP-1 is of particular interest as GLP-1 agonists are currently used for treating type 2 diabetes mellitus (Gilbert and Pratley 2020), and its pathogenesis is thought to significantly contribute to the pathogenesis of NAFLD (Tanase et al. 2020).

With the growing evidence suggesting the role of endotoxemia in the progression of NAFLD, JKB-121 has emerged as a potential therapeutic option in the treatment of NAFLD by antagonizing the toll-like receptor 4 (TLR-4), a receptor the activation of which leads to the activation of Kuppffer cells and the progression of NAFLD. In a preliminary *in vitro* study, JKB-121 showed a protective hepatic effect, likely thought to result from its anti-inflammatory action (Rivera et al. 2007). FMT is another modality focused on introducing gut microbiota, that may be concurrently associated with positive outcomes in the progression of NAFLD. Surprisingly, the premise of FMT arises from improved peripheral insulin sensitivity, as suggested by metabolic studies (Kootte et al. 2017). However, FMT studies have only been carried out on murine study samples, limiting the overall conclusions and their applicability to human subjects. Such viewpoints may be particularly susceptible to change as randomized controlled trials on humans are currently underway. Diet, through which alterations in gut microbiota have also been suggested, appears to play a significant role in disease progression. A Mediterranean diet with a high intake of legumes, vegetables and fruits has been recommended as on option that clinicians can consider when educating NAFLD patients on appropriate diets (Velasco et al. 2014). Clinicians can also take carbohydrate reduction into consideration, rather than total nutrient reduction, as a possible adjunct therapeutic intervention when treating patients with NAFLD. This is because a low-carb diet has been associated with reduced intra-hepatic triglycerides, thus raising the alarm for the possibility of improved clinical outcomes in NAFLD patients (Browning et al. 2011).

References

Food and Agricultural Organization of the United Nations and World Health Organization. Health and nutritional properties of probiotics in food including powder milk with live lactic acid bacteria. (2001). World Health Organization [online].

Adachi, R., Honma, Y., Masuno, H., Kawana, K., Shimomura, I., Yamada, S. and Makishima, M. (2005). Selective activation of vitamin D receptor by lithocholic acid acetate, a bile acid derivative. J. Lipid Res., 46(1): 46–57.

Agus, A., Clement, K. and Sokol, H. (2021). Gut microbiota-derived metabolites as central regulators in metabolic disorders. Gut, 70(6): 1174–1182.

Ahn, S. B., Jun, D. W., Kang, B. K., Lim, J. H., Lim, S. and Chung, M. J. (2019). Randomized, double-blind, placebo-controlled study of a multispecies probiotic mixture in nonalcoholic fatty liver disease. Sci. Rep., 9(1): 5688.

Alisi, A., Bedogni, G., Baviera, G., Giorgio, V., Porro, E., Paris, C., Giammaria, P., Reali, L., Anania, F. and Nobili, V. (2014). Randomised clinical trial: The beneficial effects of VSL#3 in obese children with non-alcoholic steatohepatitis. Aliment Pharmacol. Ther., 39(11): 1276–1285.

Baker, S. S., Baker, R. D., Liu, W., Nowak, N. J. and Zhu, L. (2010). Role of alcohol metabolism in non-alcoholic steatohepatitis. PLoS One, 5(3): e9570.

Begley, M., Gahan, C. G. and Hill, C. (2005). The interaction between bacteria and bile. FEMS Microbiol. Rev., 29(4): 625–651.

Bindels, L. B., Porporato, P., Dewulf, E. M., Verrax, J., Neyrinck, A. M., Martin, J. C., Scott, K. P., Buc Calderon, P., Feron, O., Muccioli, G. G., Sonveaux, P., Cani, P. D. and Delzenne, N. M. (2012). Gut microbiota-derived propionate reduces cancer cell proliferation in the liver. Br. J. Cancer, 107(8): 1337–1344.

Bomhof, M. R., Parnell, J. A., Ramay, H. R., Crotty, P., Rioux, K. P., Probert, C. S., Jayakumar, S., Raman, M. and Reimer, R. A. (2019). Histological improvement of non-alcoholic steatohepatitis with a prebiotic: A pilot clinical trial. Eur. J. Nutr., 58(4): 1735–1745.

Browning, J. D., Baker, J. A., Rogers, T., Davis, J., Satapati, S. and Burgess, S. C. (2011). Short-term weight loss and hepatic triglyceride reduction: Evidence of a metabolic advantage with dietary carbohydrate restriction. Am. J. Clin. Nutr., 93(5): 1048–1052.

Cederbaum, A. I. (2012). Alcohol metabolism. Clin. Liver Dis., 16(4): 667–685.

Chiang, J. Y. (2013). Bile acid metabolism and signaling. Compr. Physiol., 3(3): 1191–1212.

Dai, Z. L., Wu, G. and Zhu, W. Y. (2011). Amino acid metabolism in intestinal bacteria: Links between gut ecology and host health. Front. Biosci. (Landmark Ed), 16(5): 1768–1786.

Daubioul, C. A., Horsmans, Y., Lambert, P., Danse, E. and Delzenne, N. M. (2005). Effects of oligofructose on glucose and lipid metabolism in patients with nonalcoholic steatohepatitis: Results of a pilot study. Eur. J. Clin. Nutr., 59(5): 723–726.

Ding, Y., Yanagi, K., Cheng, C., Alaniz, R. C., Lee, K. and Jayaraman, A. (2019). Interactions between gut microbiota and non-alcoholic liver disease: The role of microbiota-derived metabolites. Pharmacol. Res., 141: 521–529.

Dumas, M. E., Barton, R. H., Toye, A., Cloarec, O., Blancher, C., Rothwell, A., Fearnside, J., Tatoud, R., Blanc, V., Lindon, J. C., Mitchell, S. C., Holmes, E., McCarthy, M. I., Scott, J., Gauguier, D. and Nicholson, J. K. (2006). Metabolic profiling reveals a contribution of gut microbiota to fatty liver phenotype in insulin-resistant mice. Proc. Natl. Acad. Sci. U S A, 103(33): 12511–12516.

Elshaghabee, F. M., Bockelmann, W., Meske, D., de Vrese, M., Walte, H. G., Schrezenmeir, J. and Heller, K. J. (2016). Ethanol production by selected intestinal microorganisms and lactic acid bacteria growing under different nutritional conditions. Front. Microbiol., 7: 47.

Fan, P., Li, L., Rezaei, A., Eslamfam, S., Che, D. and Ma, X. (2015). Metabolites of dietary protein and peptides by intestinal microbes and their impacts on gut. Curr. Protein Pept. Sci., 16(7): 646–654.

Ferslew, B. C., Xie, G., Johnston, C. K., Su, M., Stewart, P. W., Jia, W., Brouwer, K. L. and Barritt, A. S. t. (2015). Altered bile acid metabolome in patients with nonalcoholic steatohepatitis. Dig. Dis. Sci., 60(11): 3318–3328.

Frazier, T. H., Stocker, A. M., Kershner, N. A., Marsano, L. S. and McClain, C. J. (2011). Treatment of alcoholic liver disease. Therap. Adv. Gastroenterol., 4(1): 63–81.

Galli, A., Price, D. and Crabb, D. (1999). High-level expression of rat class I alcohol dehydrogenase is sufficient for ethanol-induced fat accumulation in transduced HeLa cells. Hepatology, 29(4): 1164–1170.

Ghosh, S. and Iacucci, M. (2021). Diverse immune effects of bovine colostrum and benefits in human health and disease. Nutrients, 13(11).

Gibson, G. R., Hutkins, R., Sanders, M. E., Prescott, S. L., Reimer, R. A., Salminen, S. J., Scott, K., Stanton, C., Swanson, K. S., Cani, P. D., Verbeke, K. and Reid, G. (2017). Expert consensus document: The International Scientific Association for Probiotics and Prebiotics (ISAPP) consensus statement on the definition and scope of prebiotics. Nat. Rev. Gastroenterol. Hepatol., 14(8): 491–502.

Gilbert, M. P. and Pratley, R. E. (2020). GLP-1 analogs and DPP-4 inhibitors in type 2 diabetes therapy: Review of head-to-head clinical trials. Front. Endocrinol. (Lausanne), 11: 178.

Gupta, H., Min, B. H., Ganesan, R., Gebru, Y. A., Sharma, S. P., Park, E., Won, S. M., Jeong, J. J., Lee, S. B., Cha, M. G., Kwon, G. H., Jeong, M. K., Hyun, J. Y., Eom, J. A., Park, H. J., Yoon, S. J., Choi, M. R., Kim, D. J. and Suk, K. T. (2022). Gut microbiome in non-alcoholic fatty liver disease: from mechanisms to therapeutic role. Biomedicines, 10(3).

Gupta, H., Youn, G. S., Shin, M. J. and Suk, K. T. (2019). Role of gut microbiota in hepatocarcinogenesis. Microorganisms, 7(5).

Harjumaki, R., Pridgeon, C. S. and Ingelman-Sundberg, M. (2021). CYP2E1 in Alcoholic and non-alcoholic liver injury. Roles of ROS, Reactive Intermediates and Lipid Overload. Int. J. Mol. Sci., 22(15).

Haussinger, D. and Schliess, F. (2007). Glutamine metabolism and signaling in the liver. Front. Biosci., 12: 371–391.

Hoyles, L., Fernandez-Real, J. M., Federici, M., Serino, M., Abbott, J., Charpentier, J., Heymes, C., Luque, J. L., Anthony, E., Barton, R. H., Chilloux, J., Myridakis, A., Martinez-Gili, L., Moreno-Navarrete, J. M., Benhamed, F., Azalbert, V., Blasco-Baque, V., Puig, J., Xifra, G., Ricart, W., Tomlinson, C., Woodbridge, M., Cardellini, M., Davato, F., Cardolini, I., Porzio, O., Gentileschi, P., Lopez, F., Foufelle, F., Butcher, S. A., Holmes, E., Nicholson, J. K., Postic, C., Burcelin, R. and Dumas, M. E. (2018). Molecular phenomics and metagenomics of hepatic steatosis in non-diabetic obese women. Nat. Med., 24(7): 1070–1080.

Hrncir, T. (2022). Gut microbiota dysbiosis: Triggers, consequences, diagnostic and therapeutic options. Microorganisms, 10(3).

Hyun, J., Han, J., Lee, C., Yoon, M. and Jung, Y. (2021). Pathophysiological Aspects of alcohol metabolism in the liver. Int. J. Mol. Sci., 22(11).

Jegatheesan, P., Beutheu, S., Ventura, G., Sarfati, G., Nubret, E., Kapel, N., Waligora-Dupriet, A. J., Bergheim, I., Cynober, L. and De-Bandt, J. P. (2016). Effect of specific amino acids on hepatic lipid metabolism in fructose-induced non-alcoholic fatty liver disease. Clin. Nutr., 35(1): 175–182.

Jiao, N., Baker, S. S., Chapa-Rodriguez, A., Liu, W., Nugent, C. A., Tsompana, M., Mastrandrea, L., Buck, M. J., Baker, R. D., Genco, R. J., Zhu, R. and Zhu, L. (2018). Suppressed hepatic bile acid signalling despite elevated production of primary and secondary bile acids in NAFLD. Gut, 67(10): 1881–1891.

Portune, K. J., Beaumount, M. Davilam A. M., Tome, D., Blachier, F. and Sanz, Y. (2016). Gut microbiota role in dietary protein metabolism and health-related outcomes: The two sides of the coin. Trends in Food Sci. Technol., 57(B): 213–232. https://doi.org/10.1016/j.tifs.2016.08.011.

Kootte, R. S., Levin, E., Salojarvi, J., Smits, L. P., Hartstra, A. V., Udayappan, S. D., Hermes, G., Bouter, K. E., Koopen, A. M., Holst, J. J., Knop, F. K., Blaak, E. E., Zhao, J., Smidt, H., Harms, A. C., Hankemeijer, T., Bergman, J., Romijn, H. A., Schaap, F. G., Olde Damink, S. W. M., Ackermans, M. T., Dallinga-Thie, G. M., Zoetendal, E., de Vos, W. M., Serlie, M. J., Stroes, E. S. G., Groen, A. K. and Nieuwdorp, M. (2017). Improvement of insulin sensitivity after lean donor feces in metabolic syndrome is driven by baseline intestinal microbiota composition. Cell. Metab., 26(4): 611–619 e616.

Lee, N. Y., Shin, M. J., Youn, G. S., Yoon, S. J., Choi, Y. R., Kim, H. S., Gupta, H., Han, S. H., Kim, B. K., Lee, D. Y., Park, T. S., Sung, H., Kim, B. Y. and Suk, K. T. (2021). Lactobacillus attenuates progression of nonalcoholic fatty liver disease by lowering cholesterol and steatosis. Clin. Mol. Hepatol., 27(1): 110–124.

Leung, H., Long, X., Ni, Y., Qian, L., Nychas, E., Siliceo, S. L., Pohl, D., Hanhineva, K., Liu, Y., Xu, A., Nielsen, H. B., Belda, E., Clement, K., Loomba, R., Li, H., Jia, W. and Panagiotou, G. (2022). Risk assessment with gut microbiome and metabolite markers in NAFLD development. Sci. Transl. Med., 14(648): eabk0855.

Leung, T. M. and Nieto, N. (2013). CYP2E1 and oxidant stress in alcoholic and non-alcoholic fatty liver disease. J. Hepatol., 58(2): 395–398.

Lieber, C. S. (1997). Ethanol metabolism, cirrhosis and alcoholism. Clin. Chim. Acta, 257(1): 59–84.

Lieber, C. S. and Schmid, R. (1961). The effect of ethanol on fatty acid metabolism; stimulation of hepatic fatty acid synthesis *in vitro*. J. Clin. Invest., 40(2): 394–399.

Lindinger, W., Taucher, J., Jordan, A., Hansel, A. and Vogel, W. (1997). Endogenous production of methanol after the consumption of fruit. Alcohol Clin. Exp. Res., 21(5): 939–943.

Liu, L., Fu, C. and Li, F. (2019). Acetate affects the process of lipid metabolism in rabbit liver, skeletal muscle and adipose tissue. Animals (Basel), 9(10).

Lucey, M. R., Mathurin, P. and Morgan, T. R. (2009). Alcoholic hepatitis. N. Engl. J. Med., 360(26): 2758–2769.

Ma, Y., Liu, X. and Wang, J. (2022). Small molecules in the big picture of gut microbiome-host cross-talk. EBioMedicine, 81: 104085.

Machado, A. S., Oliveira, J. R., Lelis, D. F., de Paula, A. M. B., Guimaraes, A. L. S., Andrade, J. M. O., Brandi, I. V. and Santos, S. H. S. (2021). Oral probiotic bifidobacterium longum supplementation improves metabolic parameters and alters the expression of the renin-angiotensin system in obese mice liver. Biol. Res. Nurs., 23(1): 100–108.

Makishima, M., Lu, T. T., Xie, W., Whitfield, G. K., Domoto, H., Evans, R. M., Haussler, M. R. and Mangelsdorf, D. J. (2002). Vitamin D receptor as an intestinal bile acid sensor. Science, 296(5571): 1313–1316.

Maldonado-Gomez, M. X., Martinez, I., Bottacini, F., O'Callaghan, A., Ventura, M., van Sinderen, D., Hillmann, B., Vangay, P., Knights, D., Hutkins, R. W. and Walter, J. (2016). Stable engraftment of Bifidobacterium longum AH1206 in the human gut depends on individualized features of the resident microbiome. Cell Host Microbe, 20(4): 515–526.

McClain, C. J., Zieve, L., Doizaki, W. M., Gilberstadt, S. and Onstad, G. R. (1980). Blood methanethiol in alcoholic liver disease with and without hepatic encephalopathy. Gut, 21(4): 318–323.

Meagher, E. A., Barry, O. P., Burke, A., Lucey, M. R., Lawson, J. A., Rokach, J. and FitzGerald, G. A. (1999). Alcohol-induced generation of lipid peroxidation products in humans. J. Clin. Invest., 104(6): 805–813.

Menni, C., Fauman, E., Erte, I., Perry, J. R., Kastenmuller, G., Shin, S. Y., Petersen, A. K., Hyde, C., Psatha, M., Ward, K. J., Yuan, W., Milburn, M., Palmer, C. N., Frayling, T. M., Trimmer, J., Bell, J. T., Gieger, C., Mohney, R. P., Brosnan, M. J., Suhre, K., Soranzo, N. and Spector, T. D. (2013). Biomarkers for type 2 diabetes and impaired fasting glucose using a nontargeted metabolomics approach. Diabetes, 62(12): 4270–4276.

Milosevic, I., Vujovic, A., Barac, A., Djelic, M., Korac, M., Radovanovic Spurnic, A., Gmizic, I., Stevanovic, O., Djordjevic, V., Lekic, N., Russo, E. and Amedei, A. (2019). Gut-liver axis, gut microbiota, and its modulation in the management of liver diseases: A review of the literature. Int. J. Mol. Sci., 20(2).

Mizrahi, M., Shabat, Y., Ben Ya'acov, A., Lalazar, G., Adar, T., Wong, V., Muller, B., Rawlin, G. and Ilan, Y. (2012). Alleviation of insulin resistance and liver damage by oral administration of Imm124-E is mediated by increased Tregs and associated with increased serum GLP-1 and adiponectin: results of a phase I/II clinical trial in NASH. J. Inflamm. Res., 5: 141–150.

Molinaro, A., Wahlstrom, A. and Marschall, H. U. (2018). Role of bile acids in metabolic control. Trends Endocrinol. Metab., 29(1): 31–41.

Mollica, M. P., Mattace Raso, G., Cavaliere, G., Trinchese, G., De Filippo, C., Aceto, S., Prisco, M., Pirozzi, C., Di Guida, F., Lama, A., Crispino, M., Tronino, D., Di Vaio, P., Berni Canani, R., Calignano, A. and Meli, R. (2017). Butyrate regulates liver mitochondrial function, efficiency, and dynamics in insulin-resistant obese mice. Diabetes, 66(5): 1405–1418.

Mouzaki, M., Wang, A. Y., Bandsma, R., Comelli, E. M., Arendt, B. M., Zhang, L., Fung, S., Fischer, S. E., McGilvray, I. G. and Allard, J. P. (2016). Bile acids and dysbiosis in non-alcoholic fatty liver disease. PLoS One, 11(5): e0151829.

Nagy, L. E. (2004). Molecular aspects of alcohol metabolism: Transcription factors involved in early ethanol-induced liver injury. Annu. Rev. Nutr., 24: 55–78.

Neis, E. P., Dejong, C. H. and Rensen, S. S. (2015). The role of microbial amino acid metabolism in host metabolism. Nutrients, 7(4): 2930–2946.

Nguyen-Lefebvre, A. T. and Horuzsko, A. (2015). Kupffer cell metabolism and function. J. Enzymol. Metab., 1(1).

Oliphant, K. and Allen-Vercoe, E. (2019). Macronutrient metabolism by the human gut microbiome: major fermentation by-products and their impact on host health. Microbiome, 7(1): 91.

Pizzorno, J. (2014). Glutathione! Integr. Med. (Encinitas), 13(1): 8–12.

Ramirez-Perez, O., Cruz-Ramon, V., Chinchilla-Lopez, P. and Mendez-Sanchez, N. (2017). The role of the gut microbiota in bile acid metabolism. Ann. Hepatol., 16(Suppl. 1: s3-105.): s15-s20.

Rehm, J., Baliunas, D., Borges, G. L., Graham, K., Irving, H., Kehoe, T., Parry, C. D., Patra, J., Popova, S., Poznyak, V., Roerecke, M., Room, R., Samokhvalov, A. V. and Taylor, B. (2010). The relation between different dimensions of alcohol consumption and burden of disease: An overview. Addiction, 105(5): 817–843.

Ridlon, J. M., Kang, D. J., Hylemon, P. B. and Bajaj, J. S. (2014). Bile acids and the gut microbiome. Curr. Opin. Gastroenterol., 30(3): 332–338.

Rivera, C. A., Adegboyega, P., van Rooijen, N., Tagalicud, A., Allman, M. and Wallace, M. (2007). Toll-like receptor-4 signaling and Kupffer cells play pivotal roles in the pathogenesis of non-alcoholic steatohepatitis. J. Hepatol., 47(4): 571–579.

Sharpton, S. R., Maraj, B., Harding-Theobald, E., Vittinghoff, E. and Terrault, N. A. (2019). Gut microbiome-targeted therapies in nonalcoholic fatty liver disease: A systematic review, meta-analysis, and meta-regression. Am. J. Clin. Nutr., 110(1): 139–149.

Sharpton, S. R., Yong, G. J. M., Terrault, N. A. and Lynch, S. V. (2019). Gut microbial metabolism and nonalcoholic fatty liver disease. Hepatol. Commun., 3(1): 29–43.

Sherriff, J. L., O'Sullivan, T. A., Properzi, C., Oddo, J. L. and Adams, L. A. (2016). Choline, its potential role in nonalcoholic fatty liver disease, and the case for human and bacterial genes. Adv. Nutr., 7(1): 5–13.

Shimazu, S. and Miklya, I. (2004). Pharmacological studies with endogenous enhancer substances: Beta-phenylethylamine, tryptamine, and their synthetic derivatives. Prog. Neuropsychopharmacol. Biol. Psychiatry, 28(3): 421–427.

Slijepcevic, D. and van de Graaf, S. F. (2017). Bile acid uptake transporters as targets for therapy. Dig. Dis., 35(3): 251–258.

Staudinger, J. L., Goodwin, B., Jones, S. A., Hawkins-Brown, D., MacKenzie, K. I., LaTour, A., Liu, Y., Klaassen, C. D., Brown, K. K., Reinhard, J., Willson, T. M., Koller, B. H. and Kliewer, S. A. (2001). The nuclear receptor PXR is a lithocholic acid sensor that protects against liver toxicity. Proc. Natl. Acad. Sci. U S A, 98(6): 3369–3374.

Szabo, G. (2015). Gut-liver axis in alcoholic liver disease. Gastroenterology, 148(1): 30–36.

Tanase, D. M., Gosav, E. M., Costea, C. F., Ciocoiu, M., Lacatusu, C. M., Maranduca, M. A., Ouatu, A. and Floria, M. (2020). The intricate relationship between type 2 diabetes Mellitus (T2DM), Insulin Resistance (IR), and Nonalcoholic Fatty Liver Disease (NAFLD). J. Diabetes Res., 2020: 3920196.

Thomas, C., Pellicciari, R., Pruzanski, M., Auwerx, J. and Schoonjans, K. (2008). Targeting bile-acid signalling for metabolic diseases. Nat. Rev. Drug Discov., 7(8): 678–693.

Torres-Fuentes, C., Schellekens, H., Dinan, T. G. and Cryan, J. F. (2017). The microbiota-gut-brain axis in obesity. Lancet Gastroenterol. Hepatol., 2(10): 747–756.

Velasco, N., Contreras, A. and Grassi, B. (2014). The Mediterranean diet, hepatic steatosis and nonalcoholic fatty liver disease. Curr. Opin. Clin. Nutr. Metab. Care, 17(5): 453–457.

Visekruna, A. and Luu, M. (2021). The role of short-chain fatty acids and bile acids in intestinal and liver function, inflammation, and carcinogenesis. Front. Cell Dev. Biol., 9: 703218.

Wang, T. J., Larson, M. G., Vasan, R. S., Cheng, S., Rhee, E. P., McCabe, E., Lewis, G. D., Fox, C. S., Jacques, P. F., Fernandez, C., O'Donnell, C. J., Carr, S. A., Mootha, V. K., Florez, J. C., Souza, A., Melander, O., Clish, C. B. and Gerszten, R. E. (2011). Metabolite profiles and the risk of developing diabetes. Nat. Med., 17(4): 448–453.

Wilfred de Alwis, N. M. and Day, C. P. (2007). Genetics of alcoholic liver disease and nonalcoholic fatty liver disease. Semin. Liver Dis., 27(1): 44–54.

Wong, V. W., G. L. Won, A. M. Chim, W. C. Chu, D. K. Yeung, K. C. Li and H. L. Chan (2013). "Treatment of nonalcoholic steatohepatitis with probiotics. A proof-of-concept study." Ann Hepatol 12(2): 256-262.

Xiao, F., Huang, Z., Li, H., Yu, J., Wang, C., Chen, S., Meng, Q., Cheng, Y., Gao, X., Li, J., Liu, Y. and Guo, F. (2011). Leucine deprivation increases hepatic insulin sensitivity via GCN2/mTOR/S6K1 and AMPK pathways. Diabetes, 60(3): 746–756.

Xue, T., Yue, L., Zhu, G., Tan, Z., Liu, H., Gan, C., Fan, C., Su, X., Xie, Y. and Ye, T. (2022). An oral phenylacrylic acid derivative suppressed hepatic stellate cell activation and ameliorated liver fibrosis by blocking TGF-beta1 signalling. Liver Int.

Yu, J. S., Youn, G. S., Choi, J., Kim, C. H., Kim, B. Y., Yang, S. J., Lee, J. H., Park, T. S., Kim, B. K., Kim, Y. B., Roh, S. W., Min, B. H., Park, H. J., Yoon, S. J., Lee, N. Y., Choi, Y. R., Kim, H. S., Gupta, H., Sung, H., Han, S. H., Suk, K. T. and Lee, D. Y. (2021). Lactobacillus lactis and Pediococcus pentosaceus-driven reprogramming of gut microbiome and metabolome ameliorates the progression of non-alcoholic fatty liver disease. Clin. Transl. Med., 11(12): e634.

Zakhari, S. (2006). Overview: how is alcohol metabolized by the body? Alcohol. Res. Health, 29(4): 245–254.

Zhu, H., Jia, Z., Misra, H. and Li, Y. R. (2012). Oxidative stress and redox signaling mechanisms of alcoholic liver disease: Updated experimental and clinical evidence. J. Dig. Dis., 13(3): 133–142.

Chapter 6

Cancer and Microbiome

Ayman H. Farran,[1] *Hanaa S. Allehaibi*[1] and *Alexandre S. Rosado*[2,*]

Introduction

What is Microbiome?

Trillions of microorganisms coexist within various parts of the human body, such as the gastrointestinal tract, lungs, mouth, skin and uterus. Although the exact number of these microorganisms is still unknown, they are incredibly diverse. The human-associated microbial communities consist of bacteria, viruses, archaea and eukaryotes. The entire collection of these microbes is called 'the human microbiota'. Conversely, 'microbiome' refers to the collection of genomes of all these microorganisms and their surrounding environment (Marchesi and Ravel 2015). The human genome comprises approximately 23,000 genes, while a bacterial genome comprises 1,000 genes on an average, depending on the microorganism. There are several distinct genotypes (subspecies) within the estimated 500–1,000 bacterial varieties present in humans, suggesting that we host bacterial genes are several hundred times greater than the total number of our human genes (Vyas and Ranganathan 2012, Bull and Plummer 2014, Valdes et al. 2018). Individual humans have host genomes that are 99.9% indistinguishable from one another (Wheeler et al. 2008). However, their gut or hand microbiomes might differ by 80–90%, promoting genetic

[1] Microbial Eco genomics and Biotechnology Laboratory (MEGB), King Abdullah University of Science and Technology.

[2] Microbial Ecogenomics and Biotechnology Laboratory (MEGB), BESE, Biological and Environmental Sciences and Engineering Division, King Abdullah University of Science and Technology (KAUST), Building 2, Level 4, Room 4217, Thuwal 23955-6900, Kingdom of Saudi Arabia.

Emails: Ayman.farran@kaust.edu.sa; Hanaa.allehaibi@kaust.edu.sa

* Corresponding author: alexandre.rosado@kaust.edu.sa

variation (Fierer et al. 2008, Turnbaugh et al. 2009a). Furthermore, the microbiome constantly changes due to many factors such as age, diet, location, probiotics and antibiotics usage, and environmental conditions (Figure 1). Thus, targeting microbiota instead of the largely stable host DNA is more adaptable and beneficial. It is crucial to understand how the diversity in microbiota impacts our health and triggers disorders. In recent years, an increased interest has been observed in the importance of the gut microbiota in the development as well as initiation of several diseases, including inflammatory bowel disorder (IBD) (Abraham and Cho 2009, Round and Mazmanian 2009), ischemic brain injury (Benakis et al. 2016), hepatic fibrosis (Bajaj et al. 2014) and obesity (Turnbaugh et al. 2009b). Our understanding of the role of the microbiome in disease etiology is continually evolving as the research continues to connect the gut microbiota to an increasing number of diseases, including cancer.

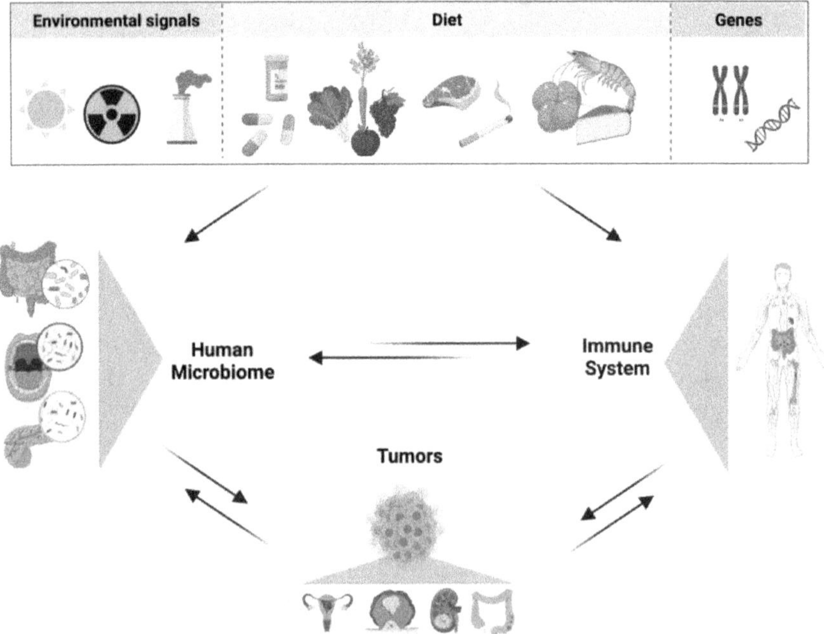

Figure 1. The relationship between the human microbiome, immune system and cancer. Both our genes and lifestyle determine the composition of our microbiome and our immune system response, which might sometimes promote cancer development and other disorders. This figure was created with BioRender.com.

Cancer and Cell Cycle

Cancer is an uncontrolled cell division caused by genetic mutations or epigenetic changes. Cancer can be hereditary, environmentally induced, or both (García-Castillo et al. 2016). The environmental risk factors of cancer are external factors (surrounding-specific), such as UV radiation, air pollution and toxic chemicals, or lifestyle factors (induvial-specific), such as age, diet, tobacco usage and alcohol consumption (Figure 1). To understand how these factors lead to a tumor, we need to be fully equipped with the knowledge of how the cell cycle runs. The cell cycle, or cell division, is a series of steps a cell undergoes to replicate into two identical daughter cells. The cell cycle is composed of interphase and mitotic (M) phases. The interphase is where the cell spends most of its time, and it consists of three stages: Gap 1 (G1), synthesis (S), and Gap 2 (G2) (Figure 2). The cell increases in terms of size and content in G1, preparing to divide. Then, it moves to the S phase, where the DNA is replicated. In the subsequent phase, G2, the cell grows further and produces the necessary proteins for mitosis. In the M phase, mitosis and cytokinesis take place. The chromosomes are aligned and segregated, and the cytoplasm is divided, forming two genetically identical daughter cells. Sometimes, cells exit G1 and enter a resting state called G_0. Cells in G_0 phase are inactive and neither dividing nor preparing to divide. G_0 phase is thought to be part of the interphase (Alberts et al. 2002, Reece et al. 2011, Raven et al. 2014).

Cell division is tightly regulated at checkpoints between the phases by a set of regulatory (positive or negative) proteins such as cyclins, cyclin-dependent kinases (CDKs), maturation-promoting factor (MPF) and anaphase-promoting complex/cyclosome (APC/C) (DiPaola 2002). Some checkpoints are along the cell cycle (Figure 2). The G1 checkpoint dictates the G1/S transition. It ensures the optimal conditions for the replication, such as size, growth factors, and DNA integrity. If these conditions are not met, the cell might go into a resting state. The G2 checkpoint (the G2/M transition) confirms the completeness of the DNA replication and scans for DNA damage before proceeding to the M phase (DiPaola 2002, Barnum and O'Connell 2014, Poon 2016). If damage is detected, the cell undergoes DNA repair. If the mutation is not repaired, then the cell goes through apoptosis (Han et al. 1995, DiPaola 2002, Barnum and O'Connell 2014, Poon 2016). This event ensures that the daughter cells do not acquire harmful or oncogenic mutations. The M checkpoint ensures the correct attachment of the sister chromatids to the spindle fibers in the metaphase before their segregation in anaphase (DiPaola 2002, Barnum and O'Connell 2014, Poon 2016). At the checkpoint, the cell checks for internal and external signals to activate the regulatory proteins that allow (or inhibit) progression to the next phase. For instance, the activity of cyclins and CDKs increases when detecting growth factors and decreases with mutations. Cancer risk factors overcome the cell cycle regulation by manipulating the signaling pathways controlling the activity of the regulatory molecules at these checkpoints.

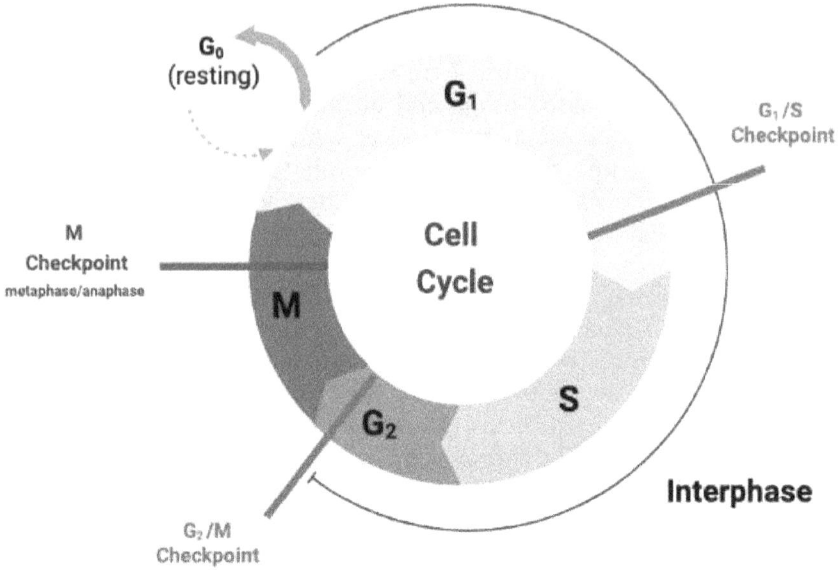

Figure 2. The cell cycle. The cell cycle consists of the interphase (G_1, S, G_2 and G_0) and the mitotic (M) phase. There are three main regulatory checkpoints to ensure that the cycle runs smoothly in the correct order: G_1/S, G_2/M and M (metaphase/anaphase). This figure was created with BioRender.com.

The Connection Between Cancer and Microbiome

There is an overlap between the factors leading to cancer and the ones influencing the microbiome. Do these factors cause cancer and microbiome modifications at the same time? Does cancer lead to microbiome changes, or does the microbiome causes cancer? Is there an order? Or is it a cycle of cause and effect? We must answer these questions to understand the relationship between our lifestyle, microbiome and cancer.

Several distinct yet complimentary molecular changes characterize tumorigenesis. Cancer cells successfully replicate and escape the immune system by activating the cell cycle and inhibiting cell cycle arrest and apoptosis. The inhibition of autophagy (a lysosome-dependent degradation mechanism for the damaged cellular components) and the activation of the metastatic pathways and angiogenesis (the formation of new blood vessels) are also considered hallmarks of cancer (Hanahan and Weinberg 2000, 2011). The microbiome's effect on these processes regulating cancer formation, progression and treatment has been poorly understood for decades, despite the prolonged research on these processes.

The connection between microbiota and cancer started being apparent with the discovery of the first oncogenic retrovirus in 1911. Peyton Rous injected a cell-free extract of chicken fibrosarcoma into other chickens, which resulted in sarcoma in the recipient chickens (Rous 1911). This experiment suggested that the tumor was transmitted by an 'agent' which was later known as Rous sarcoma virus (RSV) (Rubin 1955). This association between microbiota and cancer has been extensively studied in recent years. Zhang et al. observed changes in microbial abundance throughout the different phases of gastric cancer and described these changes as 'taxonomic biomarkers' (Zhang et al. 2021). Galeano et al. developed a single-cell RNA-sequencing tool called invasion–adhesion-directed expression sequencing (INVADEseq) to study the host-microbe interactions in colorectal and oral squamous cell carcinoma. They found a 'localized' variation per cancer type in the microbes contributing to the initiation and progression of cancer, indicating that the microbiota distribution within a tumor was niche-specific (Helmink et al. 2019, Galeano Niño et al. 2022). The intrinsic microbes residing within the tumor microenvironment (intra-tumoral microbiota or tumor-associated microbiota) communicate with one another during tumorigenesis (Nejman et al. 2020), through metabolite secretion (Pereira-Marques et al. 2019) and dysbiosis (an imbalance in the diversity of the microbial species) (Koshiol et al. 2016).

This imbalance in gut microbiota impacts both the genetic and the epigenetic pathways that regulate the development of colorectal cancer (Sun et al. 2019). The epigenetic changes control gene expression activity without changing the DNA sequence through reversible processes, such as histone modification, DNA methylation and small non-coding RNAs regulation. *Helicobacter pylori* is linked to the production of the malignant phenotype of stomach mucosa in mice. *H. pylori* infection in mouse stomach stimulates the hypermethylation of the promoter sequence of miR-490-3p, reducing its expression. miR-490-3p is a part of the microRNAs (miRNA) family, which are small non-coding RNAs that silence gene expression through the degradation of mRNAs or inhibition of translation. The downregulation of miR-490-3p reactivates the chromatin remodeler SMARCD1, which induces gastric carcinogenesis and metastasis, i.e., the migration of cancer cells to a different part of the body, away from its primary site (Shen et al. 2015). Similarly, the microbiomes in both male and female reproductive organs have significantly contributed to the development of urological and gynecological cancers in the prostate glands (Koziol et al. 2022, Shrestha et al. 2018), cervix and ovaries (Curty et al. 2020, Łaniewski et al. 2020, Nené et al. 2019). Thus, the intra-tumoral microbiome might predict susceptibility to certain malignancies (regional oncogenesis).

Moreover, cancer-associated microbiota expresses specific metabolites that disrupt the stem cell niche and the immune cells, promoting mutagenesis and cell proliferation (Abreu and Peek 2014). Therefore, gut microbiota can influence the progression of gastrointestinal cancers by producing epigenetically-relevant molecules that regulate the microbiota-sensitive epigenetic mechanisms (Rezasoltani et al. 2017, Woo and Alenghat 2022). For instance, the tumor-associated microbial communities in the gut produce short-chain fatty acids (SCFAs) and butyrate, which have anti-inflammatory properties. They inhibit histone deacetylases (HDACs), which promote the expression of antimicrobial peptides (AMPs) along with the anti-inflammatory IL-10 receptor alpha subunit (IL-10RA). SCFAs modulate the enzymatic activity of ten-eleven translocation (TET) methylcytosine dioxygenases involved in DNA methylation, while butyrate stimulates histone methylation in NF-κB1 to reduce its expression. As a result, the microbially produced SCFAs and butyrate suppress intestinal inflammation (Rezasoltani et al. 2017, Sun et al. 2019, Woo and Alenghat 2022).

In addition to pro-cancer properties, some human microbiota species exhibit immunostimulatory effects (Matson et al. 2021). Programmed cell death protein 1 (PD-1) and T-lymphocyte-associated protein 4 (CTLA-4) are cell cycle checkpoint inhibitors and T-cell receptors associated with immune inhibition (Seidel et al. 2018). Microbiota that either positively or negatively regulate PD-1 and CTLA-4 have been reported, indicating that the microbiota composition is critical in cancer immunotherapies. Therefore, they can be altered to improve the efficiency of these therapies. Microbiota modulation can be achieved by administering anti-tumor immunity and immunotherapy properties to microbial species, such as gut or fecal microbiome transplantation, probiotic bacteria, and defined bacterial consortia (Matson et al. 2021).

In this chapter, we aim to show the connection between the microbiome and the different stages of cancer, and emphasize the importance of microbiota alteration in cancer treatment.

Examples of Human Microbiomes

The Gut Microbiome

The gut microbiome refers to the microorganisms inhabiting the gastrointestinal tract. It is the largest and the most diverse human microbiome, hence labeled as a supporting organ within the digestive system. It contains over 100 trillion microbes, mostly communalistic or mutualistic, and rarely pathogenic (Savage 1977, Ogunrinola et al. 2020). The gut microbiota starts developing right from birth as the mode of delivery (vaginal delivery vs. C-section) appears to influence the microbiota composition. The infant microbiota is significantly less diverse than that of adults (Thursby and Juge 2017). After birth, the infant microbiota's diversity, composition and functionality improve as they become more exposed to nutrition, antibiotic usage, age, location and illness (Goodrich et al. 2014, Yang et al. 2020). Thus, even though studies have found microbes in the placenta, the gut microbiome is thought to be mainly environmentally acquired (Goodrich et al. 2014).

The gut microbiome interacts with host genetics and regulates core pathways, including food, vitamins and drug metabolism, short-chain fatty acids (SCFAs: acetate, propionate and butyrate) production, and immunological and inflammatory responses (Jandhyala et al. 2015, Morrison and Preston 2016). Studies on germ-free (GV) animal models that were later occupied with specific carcinogenic microbes have proposed a connection between the microbiota and the stimulation of local and systemic immunity, which contradicts the traditional belief of the microbiome's contribution in tumor induction (Vannucci et al. 2008, Mishra et al. 2021). It has been discovered that GV mice have higher colorectal and lower intestinal polyposis rates than specific pathogen-free (SPF) mice. The intestinal microbiota in SPF mice might have suppressed carcinogenesis (Mizutani et al. 1984). Furthermore, the difference in tumor occurrence could be attributed to variations in animal strains, tumor induction methodology (carcinogen type, dosage, administration route), and the organs affected (Zhan et al. 2013).

The Oral Microbiome

The oral microbiome is a term used to describe the microbial communities within the mouth. The oral cavity harbors a variety of cancer-associated microorganisms that spread to other body parts through multiple mechanisms, such as the secretion of carcinogens and inflammatory mediators, suppression of apoptosis and immune system, and the initiation of metastasis (Sun et al. 2020). Numerous studies have shown a connection between oral microbiota and the development of tumors in distant organs and oro-digestive cancers. For instance, the oral bacteria *Porphyromonas gingivalis* and *Fusobacterium nucleatum* can migrate to the colon and infiltrate the intestinal microbiome mediating gut dysbiosis, which could lead to gastrointestinal disorders and systemic inflammation (Olsen and Yamazaki 2019, Irfan et al. 2020, Mo et al. 2022) (Figure 3). Kostic et al., performed a genomic analysis using whole genome sequencing, quantitative PCR and 16S rDNA sequences in colorectal carcinomas (CRC), followed by a histological analysis. They found *Fusobacterium* species, especially *F. nucleatum*, *F. mortiferum* and *F. necrophorum*, to be noticeably enriched (Kostic et al. 2012). These findings were confirmed later by 16S rRNA sequencing and meta-analysis in CRC tissues (Drewes et al. 2017). As a result, these CRC-associated oral bacterial species could serve as cancer biomarkers (Flemer et al. 2018, Komiya et al. 2019). Even though the field is relatively young, it is already apparent that several oral plaque microbes are yet to be discovered in both primary and

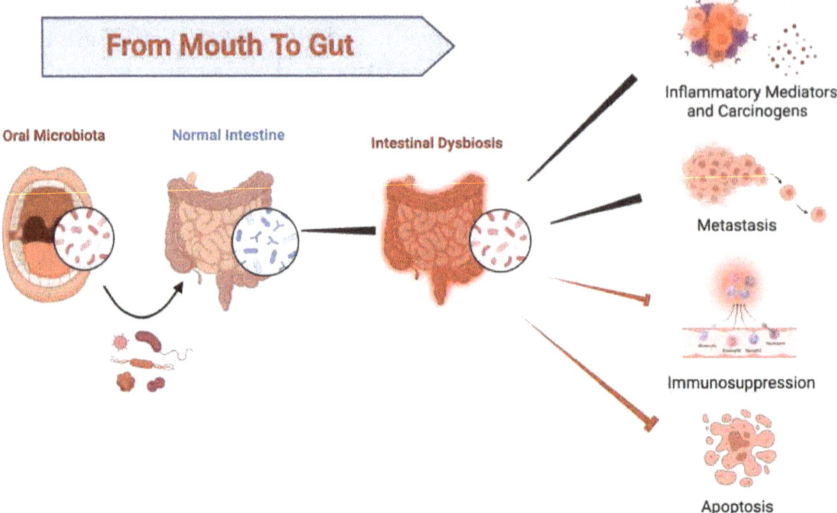

Figure 3. Oral microbiome induces gastrointestinal cancers. Some oncogenic species of the oral microbiota can migrate from the mouth to the gut and cause dysbiosis, leading to gastrointestinal disorders and dysregulation. This figure was created with BioRender.com.

metastatic cancer sites. These organisms are capable of impacting tumor inflammation (Parahitiyawa et al. 2009, Abed et al. 2016).

Microbial Inflammation in Carcinogenesis

Inflammation is a double-edged sword that can both prevent and promote tumorigenesis (Hagemann et al. 2007). Pro-inflammatory intermediates like reactive oxygen, nitrogen species, and cytokines create DNA damage within the tumor microenvironment. Also, they affect the expression and activity of DNA repair enzymes and, consequently, the genetic stability, due to the accumulation of mutations. Inflammation is well-acknowledged as a part of carcinogenesis at all stages, including cell division, metamorphosis, metastasis, penetration and angiogenesis (Figure 4). Thus, it can be considered as the eighth cancer hallmark (Hanahan and Weinberg 2000). Researchers have discovered increased cancer incidence in numerous chronic inflammatory disorders, such as inflammatory bowel disorder, pancreatitis, colitis and atrophic gastritis, in which chronic inflammation cultivates a favorable environment for tumors to thrive (Greten and Grivennikov 2019). Moreover, certain microbes induce inflammation to remodel the microenvironment for tumor initiation and then migrate along during tumor progression to sustain the inflammation in the new site (Fulbright et al. 2017, Sun et al. 2020).

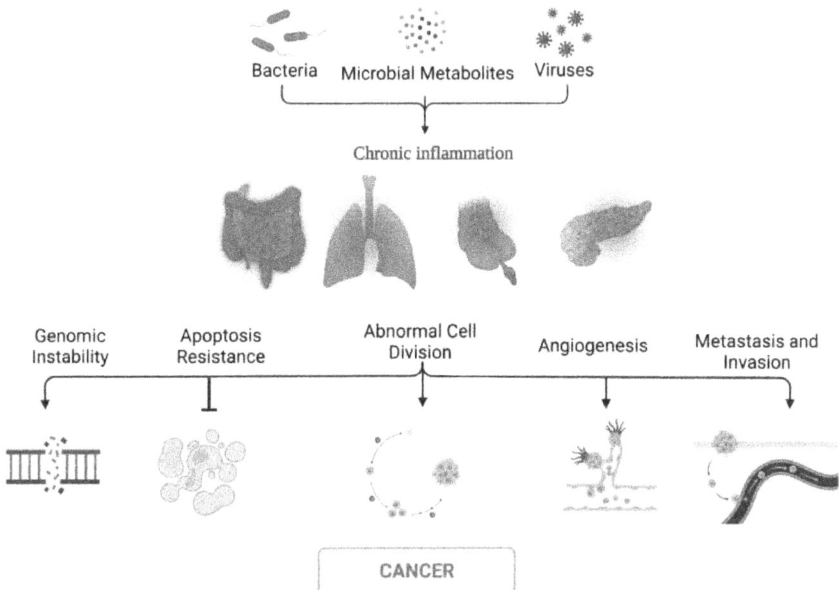

Figure 4. The role of microbial inflammation in Cancer. Pathogenic microbes induce inflammation affecting core cellular processes, leading to cancer. This figure was created with BioRender.com.

Evidence of Involvement of Microbial Inflammation in certain Cancers

Metastatic tumors contain microbial DNA traces (Marchesi et al. 2011, Gong et al. 2013). In addition, microbial infections cause about 15–16 % of cancer incidents globally (De Martel et al. 2012, Plummer et al. 2016). Bladder and gastric cancers due to *Schistosoma haematobium* (Botelho et al. 2011) and *Helicobacter pylori* (Díaz et al. 2018) infections are two examples of early malignancies linked to microbiota (Botelho et al. 2011, Díaz et al. 2018). *Fusobacterium* species have been found to be associated with colon cancer, through multiple immunosuppressive modifications, reducing T-cell activation and the cytotoxicity of natural killer (NT) cells (Gur et al. 2015, Abed et al. 2016, Abel et al. 2020). Similarly, lungs are frequently colonized by microbes. Lung microbiota typically induces regulatory T-cells (Treg) to reduce the aggressive inflammatory response due to the constant stimulation by the foreign antigens inhaled. However, it may accidentally stimulate excess inflammatory and tumorigenic reactions as well (Segal et al. 2016). Noci et al., treated mice with aerosolized antibiotics: vancomycin for Gram-positive bacteria and neomycin for Gram-negative and certain Gram-positive bacteria. They discovered a reduction in commensal bacteria and increase in T and NK cell activity, reducing the metastatic lung nodules. In short, the microbiota can enhance lung metastatic tumor development (Le Noci et al. 2018). In addition, gut microbiota can promote the development of pancreatic cancer in mice by inducing the release of the pro-tumorigenic interleukin-17 (IL-17) and reducing the number of T helper 1 (Th1) cells that kill intracellular pathogens (Sethi et al. 2018). The ineffective immunotherapy against pancreatic cancer is also a result of these immunosuppressive consequences induced by microbes (Ma et al. 2018, Sethi et al. 2019).

The Role of Mucosal Barriers

To control immunological stimulation and maintain mutualistic host-microbe interactions, the physiological isolation of intestinal microbiota from the host epithelial cells is essential (Johansson et al. 2008, Garrett 2015). Mucosal barriers protect gastric host cells from the external environment. The mucosal barrier comprises mucins secreted by goblet cells and antimicrobial peptides from Paneth cells; thus, it regulates the gut immune system. It selectively permits the passage of water and nutrients, while prohibiting bacteria and toxins (Kim and Ho 2010). When the barrier function is impaired, there is a risk of inflammation and carcinogenesis. Bacteria can then access the intestinal epithelium and release their toxic metabolites. Furthermore, they can trigger cancer development in locations distant from the colon, like the liver. Gut bacteria secrete metabolites and inflammatory molecules that can travel to the liver, where there is no known microbiome, and induce hepatocellular cancer (Schwabe and Jobin 2013). For instance, lipopolysaccharide (LPS), a component of the cell wall of Gram-negative bacteria, causes inflammation-mediated hepatocarcinogenesis in animal models (Yu et al. 2010).

The Anti-Inflammatory Properties of Gut Microbiota

In contrast to cancer-stimulating microbiota, commensal microbiota exhibits a protective behavior, protecting the host from pathogenic microbes, preventing inflammation, and maintaining the integrity of the epithelial barrier. For instance, probiotics treat gastrointestinal inflammatory conditions associated with CRC, such as inflammatory bowel disease (Kich et al. 2016). They have been shown in several *in vitro* and *in vivo* investigations to impact various carcinogenic processes, including the prevention of tumor spread and invasion, re-activation of apoptosis, and blockage of the cell cycle. The anti-inflammatory effects of probiotics lie in their ability to bind and deactivate toxins, produce anti-carcinogenic metabolites, maintain homeostasis and pH, regulate the immune system, and restore intestinal microbiota diversity and enzymatic activities. The two most prevalent probiotics in the digestive system are *Lactobacillus* and *Bifidobacterium*. *Lactobacillus* inhibits the expression of polyamines and tumor-specific antigens, reduces inflammation and DNA damage, and produces anti-oxidants and anti-angiogenetic proteins that fight cancer (Eslami et al. 2019).

Microbiome in Specific Types of Cancer

The Microbiome and Colorectal Cancer

The gut microbiota plays a significant role in colorectal cancer (CRC). In recent years, many studies have investigated the complicated interactions between the gut microbiome, tumor microbiome and the immune system throughout CRC development (Inamura 2018). It has been found that a few bacterial species enrich the stool and tumor samples of CRC patients, compared to healthy samples; these bacteria include *Escherichia coli, Bacteroides fragilis, Streptococcus gallolyticus, Enterococcus faecalis, Peptostreptococcus, Prevotella, Parvimonas, Porphyromonas* and *Fusobacterium nucleatum* (Grenham et al. 2011, Saus et al. 2019, Wong and Yu 2019). These bacterial biomarkers hold prognostic relevance in CRC diagnosis and forecasting clinical outcomes (Saus et al. 2019). However, different bacterial species have been identified in different CRC patients, indicating that CRC is controlled by a community of microbes that act synergistically (CRC-associated microbial community) instead of one specific species (Sears and Garrett 2014).

Eubiosis is when the gut microbiome in a healthy state. The integrity of the mucus layer and the epithelial barrier, balance in the gut microbiota composition, and stability between pro- and anti-inflammatory cytokines are the main characteristics of eubiosis. However, CRC-associated microbes can disturb the gut microbiota and induce a state of imbalance, dysbiosis (Al-Rashidi 2022) (Figure 5). Two prominent theories have been proposed to explain the role of CRC biomarkers in carcinogenesis. The first theory is the 'keystone-pathogen' theory (Hajishengallis et al. 2012). Keystone pathogens, such as *Porphyromonas gingivalis* and *Salmonella enterica* serovar *Typhimurium* significantly impact the microbial community, despite being a minority (Honda 2011, Maloy and Powrie 2011, Hajishengallis et al. 2012, Lamont and Hajishengallis 2015, Saus et al. 2019). They control several signaling pathways that may lead to other microbiota species creating genotoxins or other toxic byproducts (Inamura 2018). They suppress the commensal gut microbes and remodel some to be dysbiotic. Eventually, gut dysbiosis induces a systemic inflammation, leading to epithelial cell transformation and cancer (Honda 2011, Hajishengallis et al. 2012, Lamont and Hajishengallis 2015). The keystone pathogens in CRC are sometimes called 'alpha-bug'. The 'alpha-bug' hypothesis states that certain oncogenic bacterial stains or 'alpha-bugs', such as the non-enterotoxigenic strains of *Bacteroides fragilis* (NTBF), remodel the colonic microbiota to induce CRC (Sears and Pardoll 2011, Hajishengallis et al. 2012).

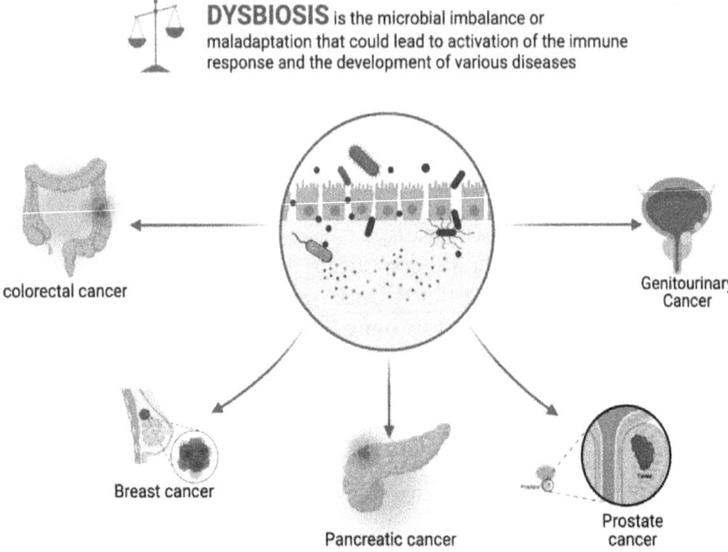

Figure 5. Gut dysbiosis is involved in various types of cancer. The figure illustrates the significant role of gut dysbiosis in the development and progression of various types of cancer. It highlights the intricate relationship between the gut microbiota and cancer, emphasizing the impact of dysbiosis on disease pathogenesis.

The other theory is the 'driver-passenger' theory, a multi-step process (Tjalsma et al. 2012). Microbes can be directly (driver) or indirectly (passenger or opportunistic) carcinogenic. For instance, *B. fragilis* and *E. faecalis* produce toxins or induce DNA damage that directly destroys gut epithelial cells (Wu et al. 2009, Goodwin et al. 2011). At the same time, others produce metabolites that activate pro-carcinogenic signaling pathways, resulting in molecular alterations. Bacterial drivers initiate CRC by triggering dysbiosis and sustained inflammation, allowing the bacterial passengers to overtake and facilitate CRC progression. This situation suggests a temporal change in the microbial composition within the tumor microenvironment at each stage of CRC (Tjalsma et al. 2012, Saus et al. 2019).

The Microbiome and Breast Cancer

Estrogen is the primary sex hormone in females. Estrogen fluctuates throughout life, from menstruation and pregnancy to menopause. It also plays a role in skin, bone and brain health. There are four significant types of estrogen: estrone, estradiol, estriol and estetrol, estradiol being the most active. It is mainly made in the ovaries and is involved in the menstrual cycle. Estrone is produced in fat tissues. Estriol and estetrol are pregnancy estrogens – estriol is made in the placenta, and estetrol in the fetal liver. All estrogens are similar in chemical structure, albeit with different functional groups; some are derivatives of others (Jones and Lopez 2006, Holinka et al. 2008, Cui et al. 2013).

Estrogen is one of the key contributors to breast cancer (BC), as 80% of BCs are estrogen-dependent (Lumachi et al. 2015). Diet, obesity, antibiotics, contraception and hormonal therapies are closely associated with changes in estrogen levels (Newman et al. 2019, Ruo et al. 2021). In addition, the gut microbiota influences estrogen metabolism and utilizes it in the production of microbial-derived metabolites that play a role in the development of BC (Rea et al. 2018, Mikó et al. 2019). Estrogen is primarily metabolized in the liver by hydroxylation and conjugation. The conjugated estrogen then enters the digestive system and becomes deconjugated into free estrogen metabolites. In the distal intestine, these metabolites are resorbed and later transported to other tissues like the breast (Yang et al. 2017). Diet alters the gut microbiota composition and enzymatic activity, resulting in changed estrogen levels. Gut microbiota possesses β-glucuronidase activity that deconjugates glucuronic acids from the conjugated estrogens, leading to reabsorption into the bloodstream as unconjugated estrogens are absorbed faster. A high-fiber diet can reduce β-glucuronidase activity and estrogen levels, hence the risk of breast cancer (Rea et al. 2018, Ruo et al. 2021).

Additionally, antibiotics may reduce the diversity of the gut microbiota composition, increasing the risk of breast disorders (Rea et al. 2018, Ruo et al. 2021). Gut dysbiosis due to antibiotics or diet can lead to obesity, a breast cancer risk factor that alters estrogen levels (Cox et al. 2014, Newman et al. 2019). On the other hand, prebiotics and probiotics, such as *Lactobacillus*, can be beneficial in preventing obesity and breast cancer as they preserve the typical microbial composition and functionality (Goldin et al. 1980, Lê et al. 1986).

Breast cancer risk is inversely correlated with the relative amounts of circulating estrogen metabolites (hydroxylated) compared to parent estrogens. Investigators have found that postmenopausal women with weaker estrogen metabolism, specifically hydroxylation, have a higher susceptibility to breast cancer (Fuhrman et al. 2012, Falk et al. 2013). Furthermore, the relationship between gut microbiome diversity and the levels of estrogens and their metabolites has been explored in various studies. The urinary ratio of hydroxylated metabolites to parent estrogens correlates to the gut microbiome diversity in healthy postmenopausal women (Fuhrman et al. 2014). The fecal estrogen levels in men and postmenopausal women show consistent results (Flores et al. 2012). A recent study showed links between dietary fibers, fecal microbiota and estrogen metabolism in postmenopausal women with breast cancer (Zengul et al. 2021). In conclusion, gut dysbiosis influences BC development through estrogen. Thus, we need to profile the gut microbiota in BC patients to identify the species involved in carcinogenesis and further investigate their disruptive mechanisms, which can be helpful in BC prevention, diagnosis or even treatment by modulation.

Microbiome-Based Cancer Therapy

Due to the dual function of bacteria in either silencing or promoting carcinogenesis, controlling the bacterial community can be helpful in cancer avoidance and therapy. In this regard, several plans for altering the configuration of the gut microbiota have been put forth:

- The oral usage of either probiotic or prebiotic (non-digestible food components incited by beneficial bacteria but not pathogens) bacteria to promote the development as well as the action of helpful intestinal bacteria.
- The simultaneous usage of probiotics and prebiotics in mixtures (known as symbiotics).
- The specific targeting of oncogenic bacteria with antibiotics (Markowiak and Ślizewska 2017, Śliżewska et al. 2020).

Strategies for Microbiome Modulation in Anti-Cancer Treatments

The markup of gut microbiota dictates the beneficial as well as harmful effects of treatments such as chemotherapy, radiation therapy and immunotherapy. Anti-cancer drugs may be used with antibiotics or economically accessible probiotic enhancements that contain valuable bacterial varieties in order to boost medical effectiveness and control adverse outcomes (Roy and Trinchieri 2017). Irinotecan (CPT-11), a camptothecin imitative, inhibits topoisomerase I as an anti-neoplastic medication. The hydrolysis of SN-38G (the inactive form of SN-38 cleansed in the liver) by bacterial β-glucuronidases in the intestinal lumen causes diarrhea with dosage restrictions (Takasuna et al. 1996). Studies have used various methods, such as antibiotics, probiotics and the careful preservation of β-glucuronidases activity, to lessen the frequency and severity of diarrhea caused by irinotecan (Mego et al. 2015). Chitapanarux et al. have demonstrated that ingesting live *Lactobacillus acidophilus* with *Bifidobacterium bifidum* reduces the intensity and occurrence of diarrhea after pelvic radiation (Chitapanarux et al. 2010). Similarly, the oral administration of *Bifidobacterium* probiotics after Ipilimumab (anti-CTLA-4) therapy decreases colitis, without affecting the drug's ability to treat tumors (Wang et al. 2018). *Fusobacterium*-enriched CRC treatment may benefit from focusing on these anaerobic Gram-negative bacteria since they reduce the anti-cancer function of Infiltrated-T and NK cells and boost cancer cell production, tumor development and metastasis in *F. nucleatum*-colonized CRC (Gur et al. 2019, Chen et al. 2020).

It should be emphasized that using antibiotics may result in an unintentional and general eradication of bacterial species (Panebianco et al. 2018). To decrease the negative impacts of traditional antibiotics, it is preferable to utilize pathogen-specific antibiotics with a limited spectrum and selective cytotoxicity for bacterial varieties (Chaurasia et al. 2016, Bhatt et al. 2017). Another piece of evidence is the increased medical reaction to ipilimumab in melanoma patients treated with anti-CTLA4 due to an increased abundance of *Bacteroides fragilis*. Therefore, the effectiveness of anti-CTLA-4 in either germ-free mice or receiving antibiotic treatment can be restored via oral delivery of *B. fragilis*, transmission of T cells sensitive to *B. fragilis*, and vaccination with *B. fragilis* polysaccharides (Vétizou et al. 2015). *Cordyceps sinensis* polysaccharides (CSP) have been proposed as prebiotics to reduce the side effects of cyclophosphamide by upregulating TLR, NF-B mechanisms and SCFAs levels, and moderating the arrangement and variety of gut microbiota, similar to how cyclophosphamide use in mice affects Th cell differentiation (Ying et al. 2020).

A recent study recommended using fecal microbiota transplantation (FMT), where the stool is transplanted from a healthy donor into the gastrointestinal tract of an immunotherapy-resistant melanoma patient. Based on reported differences in the microbiota arrangement of immunotherapy-responder and non-responder patients of various cancer kinds, numerous medical assessments have been directed to assess the effectiveness of FMT (Davar et al. 2021, Bae et al. 2022). The introduction of FMT alters the tumor microenvironment in the gut, which improves the response to immunotherapy (Baruch et al. 2021). Furthermore, FMT has shown greater efficacy in reducing *Clostridium difficile* infection because feces contain other metabolites such as proteins, bacteriophages and bile acids (Van Nood et al. 2014).

Promising Microbial-based Approaches for Cancer Therapy

Given the significance of the role of bacterial metabolites in tumorigenesis and gut regulation, they might be considered as a viable method for curing cancer. For instance, producing SCFAs and preventing polyamine synthesis and absorption show therapeutic promise in cancer treatment (Gamble et al. 2019, Tian et al. 2020). The variety of bacterial bioactive substances with anti-cancer effects was recently reviewed by Karpiński and Adamczak (Karpiński and Adamczak 2018). Notwithstanding numerous experiments on carcinogenic and potentially anti-cancer metabolites, mostly *in vitro*, the effective dosage for clinical use has not yet been demonstrated (Jaye et al. 2022). On top of that, the complicated connection between these bacterial metabolites, the tumor microenvironment, and the immune system places yet another restriction (Tian et al. 2020). The therapeutic potential of metabolites is influenced by their concentration and various variables within the tissue context. The effect of context on optimum intensity, intervention timing, and supply of metabolites to the correct location must be considered. On the other hand, stopping the production of some metabolites is advantageous since they are carcinogenic, but there are not many pharmacological techniques available to inhibit them (Descamps et al. 2019).

Traditional Medicines Have More Side Effects than Microbiome-based Therapeutics

Cytotoxic chemotherapy is the primary anti-cancer therapy for people with metastatic CRC (Hammond et al. 2016). According to Jessup et al., adjuvant chemotherapy increases the survival of stage III colon cancer patients. However, its therapeutic impact is less pronounced in patients who were black or had a high-grade illness (Jessup et al. 2005). Immunotherapy, immune checkpoint blockade in particular, is still in its early stages for treating gastrointestinal cancer in contrast to melanoma and non-small-cell lung cancer (NSCLC). As reported by Ganesh et al., anti-PD1 antibodies, pembrolizumab and nivolumab provide therapeutic benefits in patients with dMMR-MSI-H (mismatch-repair-deficient with high microsatellite instability) metastatic CRC. The antibodies are unsuccessful in most patients with metastatic CRC, who fall into the pMMR-MSI-L (proficient mismatch repair with low microsatellite instability) group. These CRC subtypes have minimal immune cell penetration and modest mutational load. Combining immune checkpoint inhibitors (ICIs) with chemotherapy and radiation may improve their clinical effectiveness by improving T-cell infiltration (Ganesh et al. 2019).

Although conservative anti-cancer therapies (chemotherapy and immunotherapy) have improved the survival chances for CRC patients, they have been associated with immune-related side effects such as colitis and diarrhea. Additionally, immune evasion and low mutational rates present difficulties for immunotherapy in treating gastrointestinal (GI) cancers (Newsome et al. 2022). On the other hand, microbiome-based treatments have fewer side effects than conventional therapies, though additional scientific research is needed. Bacillus Calmette-Guerin (BCG) is the only microbial remedy the American Food and Drug Administration (FDA) has recommended yet. Because microbial therapies cannot entirely eradicate malignancies, combining them with other conventional therapies is recommended. Also, the location of bacteria within the tumor tissue needs to be adjusted to lower the likelihood of systemic infections (Sieow et al. 2021).

Engineered Microbes as Cancer Drugs

There are many actual bacterial routes for tumor tropism, and intravenous injection frequently results in 10,000 times more germs in tumors than in similar liver, spleen and lung tissues (Roberts et al. 2014, Zheng et al. 2017). By allowing lethal payloads to be encrypted for automated delivery by tumor-homing bacteria, this attraction for tumor tissue offers a novel drug chassis and a pure passage from synthetic biology to cancer therapeutics. *Escherichia, Bifidobacterium, Listeria, Shigella, Clostridium, Lactococcus, Vibrio* and *Salmonella* species have been genetically altered and have demonstrated anticancer effects in preclinical types via intravenous, intratumor and oral delivery paths (Kramer et al. 2018). While some strategies (Figure 6) rely on the phagocytic uptake of bacteria to deliver medications intracellularly, other strategies instruct bacteria to function as 'intratumoral bioreactors', constantly producing and releasing payloads outside the cell, as a part of the settlement (Swofford et al. 2015).

Approaches to Develop Anti-cancer Proteins

Engineered bacterial lysis is an intriguing general strategy that allows the creation or release of anticancer proteins only when a specific bacterial population density is reached (Bhatt et al. 2017, Chowdhury et al. 2019). As a result, the size of bacterial colonies is drastically reduced, and systemic toxicity is avoided. Din et al. were the first to show how non-pathogenic *E. coli* and *Salmonella* could be made to lyse at a specific population density, freeing a chemokine, hemolysin, or pro-apoptotic protein, or all three, into the TME at predetermined periodic times (Din et al. 2016). Since the bacterial populace is engineered to produce growth-death-regrowth phases, the medications are administered in cycles. In order to create and distribute an antibody-fragment against CD47, which tumors can overexpress, to prevent DC phagocytosis, Chowdhury et al. used this strategy (Chowdhury et al. 2019). Interestingly, this induced a tumor-antigen-specific CD8+ T cell reaction that inhibited metastasis and generated an abscopal impact that reduced distal non-injected tumors. The method also prevented host anemias and thrombocytopenia, often with systemic CD47 antagonism, pointing to a therapeutic possibility. If we assume that intratumoral bacteria are found to be shared among different cancer kinds, then in this case, lysis circuit projects may also offer the chance to flexibly engineer patient- or tumor-specific commensal strains or multiple strains working in feedback with one another to control payload release. Given the wide range of encodable cytotoxic payloads, a fine showing of BCT clinical effectiveness with low systemic toxicity can significantly expand the toolbox for treating cancer (Sepich-Poore et al. 2021).

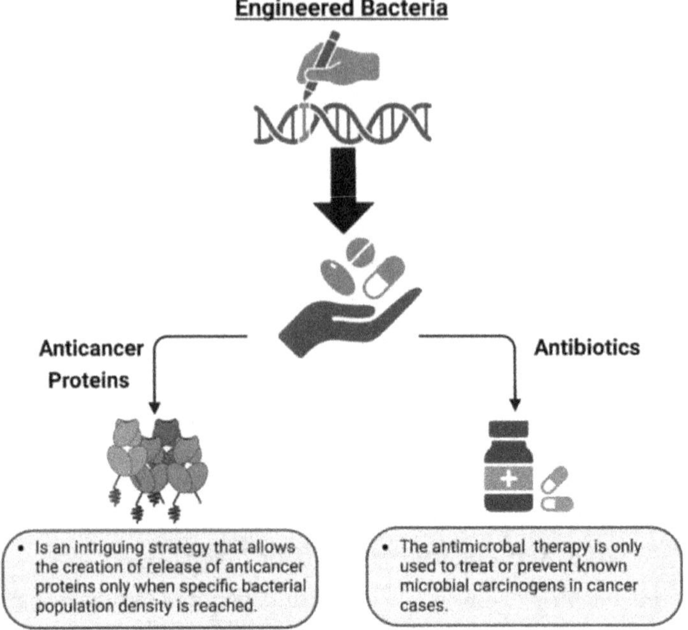

Figure 6. Approaches that are effective for cancer therapy. Two different approaches for cancer therapy: (a) selectively releasing anticancer proteins based on bacterial population density; (b) the other approach is using antimicrobial therapy to treat or prevent known microbial carcinogens in cancer cases.

Approaches to Develop Microbial Antibiotics for Cancer Therapy

Antimicrobial therapy treats or prevents known microbial carcinogens in cancer cases, including immunizing against the main human papillomavirus serotypes and hepatitis B virus to stop urogenital, cervical, head and neck, and liver cancers, as well as treating *H. pylori*-derived gastric lymphomas with triple or quadruple antibiotic treatment, giving direct-acting antivirals opposed to active Hepatitis C virus, and vaccination against known human papillomavirus serotypes (Lowy and Schiller 2017). There is anecdotal and contradictory proof regarding the use of antibiotics in solid tumors, except antibiotic-derived chemotherapies (such as doxorubicin). Reducing intratumoral microbiota may control tumor-promoting inflammatory processes, inhibit cellular proliferation, or change a tolerogenic TME into an immunogenic one, according to several studies on lung, colon and pancreatic cancers (Le Noci et al. 2018, Pushalkar et al. 2018). However, mounting clinical evidence indicates that systemic antibiotics reduce patient survival and eliminate the effectiveness of immune checkpoint inhibition (Routy et al. 2018, Huang et al. 2019). Preclinical evidence points to a delicate balance in hematologic malignancies, where either antibiotic use or gut bacterial translocation might cause leukemic development in genetically susceptible hosts (Vicente-Duenãs et al. 2020). Dietary modifications to alter the microbiota, prebiotics, postbiotics and other substances also show promise. Recent reviews on dietary influences on cancer found several epidemiological connections but few causative pathways (Murphy and Velázquez 2022). Challenges have hampered robust findings in gathering dietary data, although metabolomic information that can indicate dietary intake and associated little molecule effectors may be helpful in the times to come. Prebiotics (molecules that encourage the growth of beneficial bacteria) are currently undergoing clinical studies and show promise in boosting antitumor immunity and therapeutic response in melanoma and colon cancer (Li et al. 2020). Examples of prebiotics include resistant starch, inulin and mucin. Although the postbiotic experimental effectiveness of substances (molecules generated from microorganisms) against cancer is limited, their defined composition and manufacturing repeatability may work to their benefit (Rad et al. 2020).

Limitations and Challenges

There is a big gap between basic research and hospital clinical practices regarding microbiota and cancer. The modulation of gut microbiota in mice has shown positive results in immunotherapy studies. However, these findings have yet to be applied commercially or clinically as therapeutic approaches (Gopalakrishnan et al. 2018, Routy et al. 2018). Significant variability among human microbiome studies raises concerns regarding their reproducibility. Therefore, data concerns should be validated to facilitate translation into clinical settings. More cohort studies with large sample sizes are required to comprehend the role of intra-tumoral microorganisms in carcinogenesis, cancer expansion and cure. In addition, the correlation between the known functions of bacterial species in the gut and the 'non-bacterial' intra-tumor processes and gastrointestinal pathways linked to intestinal and colorectal cancer should be thoroughly investigated. Any overlap in function can give a perspective of the potential causes of cancer development and help redefine intra- and inter-tumor processes.

We can evaluate tumor tissues against healthy patient tissues with modern sequencing and metagenomic tools to identify novel tumor-associated microbes. We should also study whether these organisms reside within the tumor niche or are involved in the tumor's development or metastasis. However, metagenomics studies in microbial-induced cancers have generated controversial results. The controversy can be attributed to sampling type (gut microbiome in feces vs. biopsy samples), handling and preparation methods, targeted gene or variable regions, and data analysis software or pipelines (Salter et al. 2014). For instance, Jobling et al. have analyzed the fecal microbiome of 6 fecal samples by 16s rRNA gene analysis. These samples were either frozen or formalin-fixed paraffin-embedded (FFPE). The microbiome diversity varied remarkably between frozen and FFPE samples (Jobling et al. 2015). Another limitation is the minimal bacterial biomass in tumor models when using DNA extraction kits (Salter et al. 2014). Therefore, it is essential to find initiatives to regulate and standardize protocols for cancer microbiome studies, from sample collection and processing to data analysis and publication, including sampling parameters, DNA and RNA extractions, protein purification, bioinformatics, and 'multi-omics' analyses (which combine genomics, epigenomics, transcriptomics, proteomics, lipidomics and metabolomics). It is also crucial to create 'gold-standard' computational pipelines for human microbiome data processing to reduce the variability of individual microbiome studies and ensure their reproducibility, efficiency and accuracy, which is essential for their translation from basic research to clinical application. For example, the Microbiome Quality Control (MBQC) project used human fecal samples sequenced in 15 laboratories and analyzed by 9 bioinformatics tools (Sinha et al. 2017). Another option is to define general guidelines and practices to adjust to situation-specific protocols.

The Cancer Genome Atlas (TCGA) was a 12-year project with more than 11,000 patients of 33 cancer types, conducted to understand the cancer genetic landscape. It was conducted to identify the genetic characteristics of each cancer type and use them as potential therapeutic targets (The National Cancer Institute 2019). Similarly, an intensive and long-term study with several cancer patients with different cancer types can reveal the cancer microbiota landscape, such as composition, taxonomy and abundance. In addition, a meta-analysis can re-analyze and compare the existing carcinoma data from previous individual investigations in which tumor tissues, blood and stool samples were used (Eisenhofer et al. 2019). This analysis can discover global microbial drivers in cancer development, progression or treatment (Davis et al. 2018, Merritt et al. 2020). Understanding the microbiota makeup in each cancer type can enhance the microbiota modulation immunotherapy approaches in cancer treatment (McQuade et al. 2019). Interestingly, the fecal microbiota transfer (FMT) approach is more efficient and works faster than other therapies (probiotics and prebiotics) in reconstructing the intestinal microflora (Zerdan et al. 2022). Nowadays, FMT is commercially advertised as 'FMT gut restore capsules' for domestic animals.

Special multi-omics is an emerging molecular profiling tool that uses spatial information to link molecular activity and localization within a tissue. Spatial multi-omics can benefit this situation since detailed evaluations of the processes occurring within the tumor niche are required to understand the connections between cancer cells, intra-tumoral microbes and the immune system (Merritt et al. 2020, Shi et al. 2020). Organoid is another tool that can also be used to understand these interactions. It is a three-dimensional (3D) mini-organ model derived from the stem cell of the desired tissue/organ grown in a dish. Gastrointestinal organoids are derived and cocultured with different species of gut microbiota. The coculture improves the maturation and functionality of organoids by mimicking the natural microenvironment within the human tissue. This 3D model is a powerful tool for mechanistic studies that reveal the interactions between the epithelium and the surrounding environment (Min et al. 2020). In conclusion, many obstacles still exist, requiring a deeper comprehension of human-associated microbial communities in cancer patients.

Future Directions

Despite the number of studies conducted on immune-microbiome interactions in the context of cancer, there is still much more to learn. For instance, how different malignancies can selectively affect certain microbial species is still unknown. Furthermore, we are yet to understand whether cancer or the microbiome is to blame for the first disruption. Is cancer the cause of microbiome abnormalities or vice versa? Further studies on the cytokines, circulating metabolites, extracellular vesicle signaling and transportation, and bacterial quorum sensing might provide the earliest cues. The conclusions obtained should be validated in animal or human-derived preclinical models. Most cancer-associated microbial studies focus on bacteria. However, viral and fungal species need integration to get more data to fully understand the communication and coordination mechanisms within the microbiota species and the rules each class of species plays in these interactions. As a result, it would be easier to predict their sole and combined functions regarding tumor growth or anti-tumor immunity. Recent investigations have made progress in the debate over whether microbial immunomodulation occurs locally versus distantly (Jin et al. 2019, Tanoue et al. 2019).

So far, studies have shown that diet, age, gender, medical history, psychological condition and environmental exposures can influence the microbiome. In addition, risk factors for cancer, such as obesity, smoking and alcohol usage, can result in significant microbial dysbiosis. Thus, these oncogenic factors are highly likely to promote carcinogenesis through microbial dysbiosis, which emphasizes on microbiome modulation for both prevention and treatment of cancers (Sharma et al. 2020). To address this, epidemiological investigations, where patients are divided into specified populations based on their cancer type, must determine the microbial communities affected by a specific type of cancer and how this effect differs in another population with a different cancer.

The next-generation sequencing tools provide an unmatched capability to analyze the enormous variety of microbial species residing within the host and detect the oncogenic genes in both microbiota and the host. They can also provide evidence of whether these microorganisms increase the host's risk of developing particular malignancies. Gut bacteria have been shown to influence patient cohorts and preclinical model therapeutic responses in immunotherapy. In addition, these microorganisms may impact the toxicity linked to the treatment and dictate how patients respond to it. Therefore, there is an increased interest in using human-associated microbial communities as therapeutic targets to cure cancer and other disorders. However, a deep comprehension of host-microbiome interactions is required first to thoroughly realize this technique's full potential, limits and risks (Helmink et al. 2019).

Conclusions

The human microbiome and the development of tumor are connected. Therefore, examining the infectious population within the human microbiome is advantageous as a forecaster of potential disorders. Additionally, the modulation of this population might affect the efficacy of anti-cancer treatments. Despite extensive research, there is a need for more evidence regarding the exact mechanisms of cancer-inducing microbiota use. Thus, additional investigations are required to comprehend the pathways via which the microbiota causes cancer or inhibits carcinogenesis. More importantly, whether gut dysbiosis is the cause of cancer or its impact is still unknown. Finally, we emphasize the importance of standardized protocols for microbiome research, as the experimental specifics can significantly impact the outcomes. In addition, it is crucial to regulate and confirm the best sequencing methods for each purpose. Experiments should be replicated across studies and laboratories to establish credibility and avoid discrepancies among microbiome studies.

References

Abed, J., Emgård, J. E. M., Zamir, G., Faroja, M., Almogy, G., Grenov, A., Sol, A., Naor, R., Pikarsky, E., Atlan, K. A., Mellul, A., Chaushu, S., Manson, A. L., Earl, A. M., Ou, N., Brennan, C. A., Garrett, W. S. and Bachrach, G. (2016). Fap2 mediates fusobacterium nucleatum colorectal adenocarcinoma enrichment by binding to tumor-expressed

Gal-GalNAc. Cell Host & Microbe, 20(2): 215–225. https://doi.org/10.1016/J.CHOM.2016.07.006.

Abel, B., Fehlings, M., Nardin, A., Newell, E. and Yadav, M. (2020). Immuno-phenotyping of tumor-specific CD8 T cells using high-dimensional mass cytometry. The Journal of Immunology, 204(1_Supplement): 86.7–86.7. https://doi.org/10.4049/JIMMUNOL.204.SUPP.86.7.

Abraham, C. and Cho, J. (2009). Interleukin-23/Th17 pathways and inflammatory bowel disease. Inflammatory Bowel Diseases, 15(7): 1090–1100. https://doi.org/10.1002/IBD.20894.

Abreu, M. T. and Peek, R. M. (2014). Gastrointestinal malignancy and the microbiome. Gastroenterology, 146(6). https://doi.org/10.1053/J.GASTRO.2014.01.001.

Al-Rashidi, H. E. (2022). Gut microbiota and immunity relevance in eubiosis and dysbiosis. Saudi Journal of Biological Sciences, 29(3): 1628–1643. https://doi.org/10.1016/J.SJBS.2021.10.068.

Alberts, B., Johnson, A., Lewis, J., Raff, M., Roberts, K. and Walter, P. (2002). An overview of the cell cycle. *In*: Alberts, B., Johnson, A., Lewis, J., Raff, M., Roberts, K. and Walter, P. (eds.). Molecular Biology of the Cell. 4th edition. Garland Science. https://www.ncbi.nlm.nih.gov/books/NBK26869/.

Bae, J., Park, K. and Kim, Y. M. (2022). Commensal microbiota and cancer immunotherapy: harnessing commensal bacteria for cancer therapy. Immune Network, 22(1). https://doi.org/10.4110/IN.2022.22.E3.

Bajaj, J. S., Heuman, D. M., Hylemon, P. B., Sanyal, A. J., White, M. B., Monteith, P., Noble, N. A., Unser, A. B., Daita, K., Fisher, A. R., Sikaroodi, M. and Gillevet, P. M. (2014). Altered profile of human gut microbiome is associated with cirrhosis and its complications. Journal of Hepatology, 60(5): 940–947. https://doi.org/10.1016/J.JHEP.2013.12.019.

Barnum, K. J. and O'Connell, M. J. (2014). Cell cycle regulation by checkpoints. Methods in Molecular Biology (Clifton, N.J.), 1170: 29–40. https://doi.org/10.1007/978-1-4939-0888-2_2.

Baruch, E. N., Youngster, I., Ben-Betzalel, G., Ortenberg, R., Lahat, A., Katz, L., Adler, K., Dick-Necula, D., Raskin, S., Bloch, N., Rotin, D., Anafi, L., Avivi, C., Melnichenko, J., Steinberg-Silman, Y., Mamtani, R., Harati, H., Asher, N., Shapira-Frommer, R., Brosh-Nissimov, T., Eshet, Y., Ben-Simon, S., Ziv, O., Khan, M .A. W., Amit, M., Ajami, N. J., Barshack, I., Schachter, J., Wargo, J. A., Koren, O., Markel, G. and Boursi, B. (2021). Fecal microbiota transplant promotes response in immunotherapy-refractory melanoma patients. Science (New York, N.Y.), 371(6529): 602–609. https://doi.org/10.1126/SCIENCE.ABB5920.

Benakis, C., Brea, D., Caballero, S., Faraco, G., Moore, J., Murphy, M., Sita, G., Racchumi, G., Ling, L., Pamer, E. G., Iadecola, C. and Anrather, J. (2016). Commensal microbiota affects ischemic stroke outcome by regulating intestinal γδ T cells. Nature Medicine, 22(5): 516–523. https://doi.org/10.1038/NM.4068.

Bhatt, A. P., Redinbo, M. R. and Bultman, S. J. (2017). The role of the microbiome in cancer development and therapy. CA: A Cancer Journal for Clinicians, 67(4): 326. https://doi.org/10.3322/CAAC.21398.

Botelho, M. C., Oliveira, P. A., Lopes, C., Correia da Costa, J. M. and Machado, J. C. (2011). Urothelial dysplasia and inflammation induced by Schistosoma haematobium total antigen instillation in mice normal urothelium. Urologic Oncology, 29(6): 809–814. https://doi.org/10.1016/J.UROLONC.2009.09.017.

Bull, M. J. and Plummer, N. T. (2014). Part 1: The human gut microbiome in health and disease. Integrative Medicine: A Clinician's Journal, 13(6): 17. /pmc/articles/PMC4566439/.

Chaurasia, A. K., Thorat, N. D., Tandon, A., Kim, J. H., Park, S. H. and Kim, K. K. (2016). Coupling of radiofrequency with magnetic nanoparticles treatment as an alternative physical antibacterial strategy against multiple drug resistant bacteria. Scientific Reports, 6(1): 1–13. https://doi.org/10.1038/srep33662.

Chen, Y., Chen, Y., Zhang, J., Cao, P., Su, W., Deng, Y., Zhan, N., Fu, X., Huang, Y. and Dong, W. (2020). Fusobacterium nucleatum promotes metastasis in colorectal cancer by activating autophagy signaling via the upregulation of CARD3 expression. Theranostics, 10(1): 323. https://doi.org/10.7150/THNO.38870.

Chitapanarux, I., Chitapanarux, T., Traisathit, P., Kudumpee, S., Tharavichitkul, E. and Lorvidhaya, V. (2010). Randomized controlled trial of live lactobacillus acidophilus plus bifidobacterium bifidum in prophylaxis of diarrhea during radiotherapy in cervical cancer patients. Radiation Oncology (London, England), 5(1). https://doi.org/10.1186/1748-717X-5-31.

Chowdhury, S., Castro, S., Coker, C., Hinchliffe, T. E., Arpaia, N. and Danino, T. (2019). Programmable bacteria induce durable tumor regression and systemic antitumor immunity. Nature Medicine, 25(7): 1057–1063. https://doi.org/10.1038/S41591-019-0498-Z.

Cox, L. M., Yamanishi, S., Sohn, J., Alekseyenko, A. V., Leung, J. M., Cho, I., Kim, S. G., Li, H., Gao, Z., Mahana, D., Zárate Rodriguez, J. G., Rogers, A. B., Robine, N., Loke, P. and Blaser, M. J. (2014). Altering the intestinal microbiota during a critical developmental window has lasting metabolic consequences. Cell, 158(4): 705–721. https://doi.org/10.1016/J.CELL.2014.05.052.

Cui, J., Shen, Y. and Li, R. (2013). Estrogen synthesis and signaling pathways during ageing: from periphery to brain. Trends in Molecular Medicine, 19(3): 197. https://doi.org/10.1016/J.MOLMED.2012.12.007.

Curty, G., de Carvalho, P. S. and Soares, M. A. (2020). The role of the cervicovaginal microbiome on the genesis and as a biomarker of premalignant cervical intraepithelial neoplasia and invasive cervical cancer. International Journal of Molecular Sciences, 21(1). https://doi.org/10.3390/IJMS21010222.

Davar, D., Dzutsev, A. K., McCulloch, J. A., Rodrigues, R. R., Chauvin, J. M., Morrison, R. M., Deblasio, R. N., Menna, C., Ding, Q., Pagliano, O., Zidi, B., Zhang, S., Badger, J. H., Vetizou, M., Cole, A. M., Fernandes, M. R., Prescott, S., Costa, R. G. F., Balaji, A. K., Morgun, A., Vujkovic-Cvijin, I., Wang, H., Borhani, A. A., Schwartz, M. B., Dubner, H. M., Ernst, S. J., Rose, A., Najjar, Y. G., Belkaid, Y., Kirkwood, J. M., Trinchieri, G. and Zarour, H. M. (2021). Fecal microbiota transplant overcomes resistance to anti-PD-1 therapy in melanoma patients. Science (New York, N.Y.), 371(6529): 595–602. https://doi.org/10.1126/SCIENCE.ABF3363.

Davis, N. M., Proctor, Di. M., Holmes, S. P., Relman, D. A. and Callahan, B. J. (2018). Simple statistical identification and removal of contaminant sequences in marker-gene and metagenomics data. Microbiome, 6(1): 1–14. https://doi.org/10.1186/S40168-018-0605-2/FIGURES/6.

De Martel, C., Ferlay, J., Franceschi, S., Vignat, J., Bray, F., Forman, D. and Plummer, M. (2012). Global burden of cancers attributable to infections in 2008: A review and synthetic analysis. The Lancet Oncology, 13(6): 607–615. https://doi.org/10.1016/S1470-2045(12)70137-7.

Descamps, H. C., Herrmann, B., Wiredu, D. and Thaiss, C. A. (2019). The path toward using microbial metabolites as therapies. Ebiomedicine, 44: 747–754. https://doi.org/10.1016/J.EBIOM.2019.05.063.

Díaz, P., Valderrama, M. V., Bravo, J. and Quest, A. F. G. (2018). Helicobacter pylori and gastric cancer: Adaptive cellular mechanisms involved in disease progression. Frontiers in Microbiology, 9(JAN). https://doi.org/10.3389/FMICB.2018.00005.

Din, M. O., Danino, T., Prindle, A., Skalak, M., Selimkhanov, J., Allen, K., Julio, E., Atolia, E., Tsimring, L. S., Bhatia, S. N. and Hasty, J. (2016). Synchronized cycles of bacterial lysis for in vivo delivery. Nature, 536(7614): 81–85. https://doi.org/10.1038/nature18930.

DiPaola, R. S. (2002). To arrest or not to G(2)-M Cell-cycle arrest: Commentary re: A. K. Tyagi et al. Silibinin strongly synergizes human prostate carcinoma DU145 cells to doxorubicin-induced growth inhibition, G(2)-M arrest, and apoptosis, 8(11): 3512–3519. https://pubmed.ncbi.nlm.nih.gov/12429616/.

Drewes, J. L., White, J. R., Dejea, C. M., Fathi, P., Iyadorai, T., Vadivelu, J., Roslani, A. C., Wick, E. C., Mongodin, E. F., Loke, M. F., Thulasi, K., Gan, H. M., Goh, K. L., Chong, H. Y., Kumar, S., Wanyiri, J. W. and Sears, C. L. (2017). High-resolution bacterial 16S rRNA gene profile meta-analysis and biofilm status reveal common colorectal cancer consortia. NPJ Biofilms and Microbiomes, 3(1). https://doi.org/10.1038/S41522-017-0040-3.

Eisenhofer, R., Minich, J. J., Marotz, C., Cooper, A., Knight, R. and Weyrich, L. S. (2019). Contamination in low microbial biomass microbiome studies: Issues and recommendations. Trends in Microbiology, 27(2): 105–117. https://doi.org/10.1016/J.TIM.2018.11.003.

Eslami, M., Yousefi, B., Kokhaei, P., Hemati, M., Nejad, Z. R., Arabkari, V. and Namdar, A. (2019). Importance of probiotics in the prevention and treatment of colorectal cancer. Journal of Cellular Physiology, 234(10): 17127–17143. https://doi.org/10.1002/JCP.28473.

Falk, R. T., Brinton, L. A., Dorgan, J. F., Fuhrman, B. J., Veenstra, T. D., Xu, X. and Gierach, G. L. (2013). Relationship of serum estrogens and estrogen metabolites to postmenopausal breast cancer risk: A nested case-control study. Breast Cancer Research: BCR, 15(2): R34. https://doi.org/10.1186/BCR3416/FIGURES/8.

Fierer, N., Hamady, M., Lauber, C. L. and Knight, R. (2008). The influence of sex, handedness, and washing on the diversity of hand surface bacteria. Proceedings of the National Academy of Sciences of the United States of America, 105(46): 17994–17999. https://doi.org/10.1073/PNAS.0807920105.

Flemer, B., Warren, R. D., Barrett, M. P., Cisek, K., Das, A., Jeffery, I. B., Hurley, E., O'Riordain, M., Shanahan, F. and O'Toole, P. W. (2018). The oral microbiota in colorectal cancer is distinctive and predictive. Gut, 67(8): 1454–1463. https://doi.org/10.1136/GUTJNL-2017-314814.

Flores, R., Shi, J., Fuhrman, B., Xu, X., Veenstra, T. D., Gail, M. H., Gajer, P., Ravel, J. and Goedert, J. J. (2012). Fecal microbial determinants of fecal and systemic estrogens and estrogen metabolites: A cross-sectional study. Journal of Translational Medicine, 10(1). https://doi.org/10.1186/1479-5876-10-253.

Fuhrman, B. J., Feigelson, H. S., Flores, R., Gail, M. H., Xu, X., Ravel, J. and Goedert, J. J. (2014). Associations of the fecal microbiome with urinary estrogens and estrogen metabolites in postmenopausal women. The Journal of Clinical Endocrinology and Metabolism, 99(12): 4632–4640. https://doi.org/10.1210/JC.2014-2222.

Fuhrman, B. J., Schairer, C., Gail, M. H., Boyd-Morin, J., Xu, X., Sue, L. Y., Buys, S. S., Isaacs, C., Keefer, L. K., Veenstra, T. D., Berg, C. D., Hoover, R. N. and Ziegler, R. G. (2012). Estrogen metabolism and risk of breast cancer in postmenopausal women. Journal of the National Cancer Institute, 104(4): 326–339. https://doi.org/10.1093/JNCI/DJR531.

Fulbright, L. E., Ellermann, M. and Arthur, J. C. (2017). The microbiome and the hallmarks of cancer. PLoS Pathogens, 13(9). https://doi.org/10.1371/JOURNAL.PPAT.1006480.

Galeano Niño, J. L., Wu, H., LaCourse, K. D., Kempchinsky, A. G., Baryiames, A., Barber, B., Futran, N., Houlton, J., Sather, C., Sicinska, E., Taylor, A., Minot, S. S., Johnston, C. D. and Bullman, S. (2022). Effect of the intratumoral microbiota on spatial and cellular heterogeneity in cancer. Nature, 611(7937): 810. https://doi.org/10.1038/S41586-022-05435-0.

Gamble, L. D., Purgato, S., Murray, J., Xiao, L., Yu, D. M. T., Hanssen, K. M., Giorgi, F. M., Carter, D. R., Gifford, A. J., Valli, E., Milazzo, G., Kamili, A., Mayoh, C., Liu, B., Eden, G., Sarraf, S., Allan, S., Giacomo, S. Di, Flemming, C. L., Russell, A.J., Cheung, B. B., Oberthuer, A., London, W. B., Fischer, M., Trahair, T. N., Fletcher, J. I., Marshall, G.

M., Ziegler, D. S., Hogarty, M. D., Burns, M. R., Perini, G., Norris, M. D. and Haber, M. (2019). Inhibition of polyamine synthesis and uptake reduces tumor progression and prolongs survival in mouse models of neuroblastoma. Science Translational Medicine, 11(477). https://doi.org/10.1126/SCITRANSLMED.AAU1099.

Ganesh, K., Stadler, Z. K., Cercek, A., Mendelsohn, R. B., Shia, J., Segal, N. H. and Diaz, L. A. (2019). Immunotherapy in colorectal cancer: Rationale, challenges and potential. Nature Reviews. Gastroenterology & Hepatology, 16(6): 361–375. https://doi.org/10.1038/S41575-019-0126-X.

García-Castillo, V., Sanhueza, E., McNerney, E., Onate, S. A. and García, A. (2016). Microbiota dysbiosis: A new piece in the understanding of the carcinogenesis puzzle. Journal of Medical Microbiology, 65(12): 1347–1362. https://doi.org/10.1099/JMM.0.000371.

Garrett, W. S. (2015). Cancer and the microbiota. Science (New York, N.Y.), 348(6230): 80–86. https://doi.org/10.1126/SCIENCE.AAA4972.

Goldin, B. R., Swenson, L., Dwyer, J., Sexton, M. and Gorbach, S. L. (1980). Effect of diet and Lactobacillus acidophilus supplements on human fecal bacterial enzymes. Journal of the National Cancer Institute, 64(2): 255–261. https://doi.org/10.1093/JNCI/64.2.255.

Gong, H. L., Shi, Y., Zhou, L., Wu, C. P., Cao, P. Y., Tao, L., Xu, C., Hou, D. S. and Wang, Y. Z. (2013). The composition of microbiome in larynx and the throat biodiversity between laryngeal squamous cell carcinoma patients and control population. PloS One, 8(6). https://doi.org/10.1371/JOURNAL.PONE.0066476.

Goodrich, J. K., Waters, J. L., Poole, A. C., Sutter, J. L., Koren, O., Blekhman, R., Beaumont, M., Van Treuren, W., Knight, R., Bell, J. T., Spector, T. D., Clark, A. G. and Ley, R. E. (2014). Human genetics shape the gut microbiome. Cell, 159(4): 789. https://doi.org/10.1016/J.CELL.2014.09.053.

Goodwin, A. C., Destefano Shields, C. E., Wu, S., Huso, D. L., Wu, X. Q., Murray-Stewart, T. R., Hacker-Prietz, A., Rabizadeh, S., Woster, P. M., Sears, C. L. and Casero, R. A. (2011). Polyamine catabolism contributes to enterotoxigenic Bacteroides fragilis-induced colon tumorigenesis. Proceedings of the National Academy of Sciences of the United States of America, 108(37): 15354–15359. https://doi.org/10.1073/PNAS.1010203108.

Gopalakrishnan, V., Spencer, C. N., Nezi, L., Reuben, A., Andrews, M. C., Karpinets, T. V., Prieto, P. A., Vicente, D., Hoffman, K., Wei, S. C., Cogdill, A. P., Zhao, L., Hudgens, C. W., Hutchinson, D. S., Manzo, T., Petaccia De Macedo, M., Cotechini, T., Kumar, T., Chen, W. S., Reddy, S. M., Szczepaniak Sloane, R., Galloway-Pena, J., Jiang, H., Chen, P.L., Shpall, E. J., Rezvani, K., Alousi, A. M., Chemaly, R .F., Shelburne, S., Vence, L. M., Okhuysen, P. C., Jensen, V. B., Swennes, A. G., McAllister, F., Marcelo Riquelme Sanchez, E., Zhang, Y., Le Chatelier, E., Zitvogel, L., Pons, N., Austin-Breneman, J. L., Haydu, L. E., Burton, E. M., Gardner, J. M., Sirmans, E., Hu, J., Lazar, A. J., Tsujikawa, T., Diab, A., Tawbi, H., Glitza, I. C., Hwu, W. J., Patel, S. P., Woodman, S. E., Amaria, R. N., Davies, M .A., Gershenwald, J. E., Hwu, P., Lee, J. E., Zhang, J., Coussens, L. M., Cooper, Z. A., Futreal, P. A., Daniel, C. R., Ajami, N. J., Petrosino, J. F., Tetzlaff, M. T., Sharma, P., Allison, J. P., Jenq, R. R. and Wargo, J. A. (2018). Gut microbiome modulates response to anti-PD-1 immunotherapy in melanoma patients. Science (New York, N.Y.), 359(6371): 97–103. https://doi.org/10.1126/SCIENCE.AAN4236.

Grenham, S., Clarke, G., Cryan, J. F. and Dinan, T. G. (2011). Brain-gut-microbe communication in health and disease. Frontiers in Physiology, 2 DEC: 94. https://doi.org/10.3389/FPHYS.2011.00094/BIBTEX.

Greten, F. R. and Grivennikov, S. I. (2019). Inflammation and cancer: Triggers, mechanisms, and consequences. Immunity, 51(1): 27–41. https://doi.org/10.1016/J.IMMUNI.2019.06.025.

Gur, C., Ibrahim, Y., Isaacson, B., Yamin, R., Abed, J., Gamliel, M., Enk, J., Bar-On, Y., Stanietsky-Kaynan, N., Coppenhagen-Glazer, S., Shussman, N., Almogy, G., Cuapio, A., Hofer, E., Mevorach, D., Tabib, A., Ortenberg, R., Markel, G., Miklić, K., Jonjic, S., Brennan, C. A., Garrett, W. S., Bachrach, G. and Mandelboim, O. (2015). Binding of the

Fap2 protein of Fusobacterium nucleatum to human inhibitory receptor TIGIT protects tumors from immune cell attack. Immunity, 42(2): 344–355. https://doi.org/10.1016/J.IMMUNI.2015.01.010.

Gur, C., Maalouf, N., Shhadeh, A., Berhani, O., Singer, B. B., Bachrach, G. and Mandelboim, O. (2019). Fusobacterium nucleatum supresses anti-tumor immunity by activating CEACAM1. Oncoimmunology, 8(6). https://doi.org/10.1080/2162402X.2019.1581531.

Hagemann, T., Balkwill, F. and Lawrence, T. (2007). Inflammation and cancer: A double-edged sword. Cancer Cell, 12(4): 300. https://doi.org/10.1016/J.CCR.2007.10.005.

Hajishengallis, G., Darveau, R. P. and Curtis, M. A. (2012). The keystone pathogen hypothesis. Nature reviews. Microbiology, 10(10): 717. https://doi.org/10.1038/NRMICRO2873.

Hammond, W. A., Swaika, A. and Mody, K. (2016). Pharmacologic resistance in colorectal cancer: A review. Therapeutic Advances in Medical Oncology, 8(1): 57. https://doi.org/10.1177/1758834015614530.

Han, Z., Chatterjee, D., He, D. M., Early, J., Pantazis, P., Wyche, J. H. and Hendrickson, E. A. (1995). Evidence for a G2 checkpoint in p53-independent apoptosis induction by X-irradiation. Molecular and Cellular Biology, 15(11): 5849–5857. https://doi.org/10.1128/MCB.15.11.5849.

Hanahan, D. and Weinberg, R. A. (2000). The hallmarks of cancer. Cell, 100(1): 57–70. https://doi.org/10.1016/S0092-8674(00)81683-9.

Hanahan, D. and Weinberg, R. A. (2011). Hallmarks of cancer: The next generation. Cell, 144(5): 646–674. https://doi.org/10.1016/J.CELL.2011.02.013.

Helmink, B. A., Khan, M. A. W., Hermann, A., Gopalakrishnan, V. and Wargo, J. A. (2019). The microbiome, cancer, and cancer therapy. Nature Medicine, 25(3): 377–388. https://doi.org/10.1038/S41591-019-0377-7.

Holinka, C. F., Diczfalusy, E. and Coelingh Bennink, H. J. T. (2008). Estetrol: A unique steroid in human pregnancy. The Journal of Steroid Biochemistry and Molecular Biology, 110(1-2): 138–143. https://doi.org/10.1016/J.JSBMB.2008.03.027.

Honda, K. (2011). Porphyromonas gingivalis sinks teeth into the oral microbiota and periodontal disease. Cell Host & Microbe, 10(5): 423–425. https://doi.org/10.1016/J.CHOM.2011.10.008.

Huang, X. Z., Gao, P., Song, Y. X., Xu, Y., Sun, J. X., Chen, X. W., Zhao, J. H. and Wang, Z. N. (2019). Antibiotic use and the efficacy of immune checkpoint inhibitors in cancer patients: a pooled analysis of 2740 cancer patients. Oncoimmunology, 8(12). https://doi.org/10.1080/2162402X.2019.1665973.

Inamura, K. (2018). Colorectal cancers: An update on their molecular pathology. Cancers (Basel), 10(1): 26. doi: 10.3390/cancers10010026. PMID: 29361689; PMCID: PMC5789376.

Irfan, M., Delgado, R. Z. R. and Frias-Lopez, J. (2020). The oral microbiome and cancer. Frontiers in Immunology, 11. https://doi.org/10.3389/FIMMU.2020.591088.

Jandhyala, S. M., Talukdar, R., Subramanyam, C., Vuyyuru, H., Sasikala, M. and Reddy, D. N. (2015). Role of the normal gut microbiota. World Journal of Gastroenterology: WJG, 21(29): 8787. https://doi.org/10.3748/WJG.V21.I29.8787.

Jaye, K., Li, C. G., Chang, D. and Bhuyan, D. J. (2022). The role of key gut microbial metabolites in the development and treatment of cancer. Gut Microbes, 14(1). https://doi.org/10.1080/19490976.2022.2038865.

Jessup, J. M., Stewart, A., Greene, F. L. and Minsky, B. D. (2005). Adjuvant chemotherapy for stage III colon cancer: implications of race/ethnicity, age, and differentiation. JAMA, 294(21): 1092–1094. https://doi.org/10.1001/JAMA.294.21.2703.

Jin, C., Lagoudas, G. K., Zhao, C., Bullman, S., Bhutkar, A., Hu, B., Ameh, S., Sandel, D., Liang, X. S., Mazzilli, S., Whary, M. T., Meyerson, M., Germain, R., Blainey, P. C., Fox, J. G. and Jacks, T. (2019). Commensal microbiota promote lung cancer development via $\gamma\delta$ T cells. Cell, 176(5): 998–1013.e16. https://doi.org/10.1016/J.CELL.2018.12.040.

Jobling, I., Taylor, M., Young, C., Wood, H. and Quirke, P. (2015). Investigating the Faecal Microbiome in Formalin Fixed Paraffin Embedded (FFPE) Material. The Journal of Pathology. Dublin Pathology 2015. 8th Joint Meeting of the British Division of the International Academy of Pathology and the Pathological Society of Great Britain & Ireland, 237: S1–S52. https://doi.org/10.1002/PATH.4631.

Johansson, M. E. V., Phillipson, M., Petersson, J., Velcich, A., Holm, L. and Hansson, G. C. (2008). The inner of the two Muc2 mucin-dependent mucus layers in colon is devoid of bacteria. Proceedings of the National Academy of Sciences of the United States of America, 105(39): 15064–15069. https://doi.org/10.1073/PNAS.0803124105.

Jones, R. E., Richard E. and Lopez, K. H. (2006). Human Reproductive Biology. Elsevier Academic Press.

Karpiński, T. M. and Adamczak, A. (2018). Anticancer activity of bacterial proteins and peptides. Pharmaceutics, 10(2). https://doi.org/10.3390/PHARMACEUTICS10020054.

Kich, D. M., Vincenzi, A., Majolo, F., Volken de Souza, C. F. and Goettert, M. I. (2016). Probiotic: Effectiveness nutrition in cancer treatment and prevention. Nutricion Hospitalaria, 33(6): 1430–1437. https://doi.org/10.20960/NH.806.

Kim, Y. S. and Ho, S. B. (2010). Intestinal goblet cells and mucins in health and disease: Recent insights and progress. Current Gastroenterology Reports, 12(5): 319–330. https://doi.org/10.1007/S11894-010-0131-2.

Komiya, Y., Shimomura, Y., Higurashi, T., Sugi, Y., Arimoto, J., Umezawa, S., Uchiyama, S., Matsumoto, M. and Nakajima, A. (2019). Patients with colorectal cancer have identical strains of Fusobacterium nucleatum in their colorectal cancer and oral cavity. Gut, 68(7): 1335–1337. https://doi.org/10.1136/GUTJNL-2018-316661.

Koshiol, J., Wozniak, A., Cook, P., Adaniel, C., Acevedo, J., Azócar, L., Hsing, A. W., Roa, J. C., Pasetti, M. F., Miquel, J. F., Levine, M. M., Ferreccio, C., Aguayo, C. G., Baez, S., Díaz, A., Molina, H., Miranda, C., Castillo, C., Tello, A., Durán, G., Delgado, C. P., Quevedo, R., Pineda, S., la Barra, T., Reyes, C., Alegría, C., Aguayo, C., Losada, H., Arraya, J. C., Bellolio, E., Tapia, O., López, J., Medina, K., Barraza, P., Catalán, S., Riquelme, P., Órdenes, L., Garcés, R., Duarte, C. and Hildesheim, A. (2016). Salmonella enterica serovar Typhi and gallbladder cancer: a case–control study and meta-analysis. Cancer Medicine, 5(11): 3310. https://doi.org/10.1002/CAM4.915.

Kostic, A. D., Gevers, D., Pedamallu, C. S., Michaud, M., Duke, F., Earl, A. M., Ojesina, A. I., Jung, J., Bass, A. J., Tabernero, J., Baselga, J., Liu, C., Shivdasani, R. A., Ogino, S., Birren, B. W., Huttenhower, C., Garrett, W. S. and Meyerson, M. (2012). Genomic analysis identifies association of Fusobacterium with colorectal carcinoma. Genome Research, 22(2): 292. https://doi.org/10.1101/GR.126573.111.

Koziol, J. H., Sheets, T., Wickware, C. L. and Johnson, T. A. (2022). Composition and diversity of the seminal microbiota in bulls and its association with semen parameters. Theriogenology, 182: 17–25. https://doi.org/10.1016/J.THERIOGENOLOGY.2022.01.029.

Kramer, M. G., Masner, M., Ferreira, F. A. and Hoffman, R. M. (2018). Bacterial therapy of cancer: Promises, limitations, and insights for future directions. Frontiers in Microbiology, 9(JAN). https://doi.org/10.3389/FMICB.2018.00016.

Lamont, R. J. and Hajishengallis, G. (2015). Polymicrobial synergy and dysbiosis in inflammatory disease. Trends in Molecular Medicine, 21(3): 172–183. https://doi.org/10.1016/J.MOLMED.2014.11.004.

Łaniewski, P., Ilhan, Z. E. and Herbst-Kralovetz, M. M. (2020). The microbiome and gynaecological cancer development, prevention and therapy. Nature Reviews. Urology, 17(4). https://doi.org/10.1038/S41585-020-0286-Z.

Lê, M. G., Moulton, L. H., Hill, C. and Kramar, A. (1986). Consumption of dairy produce and alcohol in a case-control study of breast cancer. Journal of the National Cancer Institute, 77(3): 633–636. https://doi.org/10.1093/JNCI/77.3.633.

Le Noci, V., Guglielmetti, S., Arioli, S., Camisaschi, C., Bianchi, F., Sommariva, M., Storti, C., Triulzi, T., Castelli, C., Balsari, A., Tagliabue, E. and Sfondrini, L. (2018). Modulation of pulmonary microbiota by antibiotic or probiotic aerosol therapy: A strategy to promote immunosurveillance against lung metastases. Cell Reports, 24(13): 3528–3538. https://doi.org/10.1016/J.CELREP.2018.08.090.

Li, Y., Elmén, L., Segota, I., Xian, Y., Tinoco, R., Feng, Y., Fujita, Y., Segura Muñoz, R. R., Schmaltz, R., Bradley, L. M., Ramer-Tait, A., Zarecki, R., Long, T., Peterson, S. N. and Ronai, Z. A. (2020). Prebiotic-induced anti-tumor immunity attenuates tumor growth. Cell Reports, 30(6): 1753. https://doi.org/10.1016/J.CELREP.2020.01.035.

Lowy, D. R. and Schiller, J. T. (2017). Preventing cancer and other diseases caused by human papillomavirus infection: 2017 lasker-debakey clinical research award. JAMA, 318(10): 901. https://doi.org/10.1001/JAMA.2017.11706.

Lumachi, F., Santeufemia, D. A. and Basso, S. M. (2015). Current medical treatment of estrogen receptor-positive breast cancer. World Journal of Biological Chemistry, 6(3): 231. https://doi.org/10.4331/WJBC.V6.I3.231.

Ma, C., Han, M., Heinrich, B., Fu, Q., Zhang, Q., Sandhu, M., Agdashian, D., Terabe, M., Berzofsky, J. A., Fako, V., Ritz, T., Longerich, T., Theriot, C. M., McCulloch, J. A., Roy, S., Yuan, W., Thovarai, V., Sen, S. K., Ruchirawat, M., Korangy, F., Wang, X. W., Trinchieri, G. and Greten, T. F. (2018). Gut microbiome-mediated bile acid metabolism regulates liver cancer via NKT cells. Science (New York, N.Y.), 360(6391). https://doi.org/10.1126/SCIENCE.AAN5931.

Maloy, K. J. and Powrie, F. (2011). Intestinal homeostasis and its breakdown in inflammatory bowel disease. Nature, 474(7351): 298–306. https://doi.org/10.1038/NATURE10208.

Marchesi, J. R., Dutilh, B. E., Hall, N., Peters, W. H. M., Roelofs, R., Boleij, A. and Tjalsma, H. (2011). Towards the human colorectal cancer microbiome. PloS One, 6(5). https://doi.org/10.1371/JOURNAL.PONE.0020447.

Marchesi, J. R. and Ravel, J. (2015). The vocabulary of microbiome research: A proposal. Microbiome, 3(1). https://doi.org/10.1186/S40168-015-0094-5.

Markowiak, P. and Ślizewska, K. (2017). Effects of probiotics, prebiotics, and synbiotics on human health. Nutrients, 9(9). https://doi.org/10.3390/NU9091021.

Matson, V., Chervin, C. S. and Gajewski, T. F. (2021). Cancer and the microbiome-influence of the commensal microbiota on cancer, immune responses, and immunotherapy. Gastroenterology, 160(2): 600–613. https://doi.org/10.1053/J.GASTRO.2020.11.041.

McQuade, J. L., Daniel, C. R., Helmink, B. A. and Wargo, J. A. (2019). Modulating the microbiome to improve therapeutic response in cancer. The Lancet. Oncology, 20(2): e77–e91. https://doi.org/10.1016/S1470-2045(18)30952-5.

Mego, M., Chovanec, J., Vochyanova-Andrezalova, I., Konkolovsky, P., Mikulova, M., Reckova, M., Miskovska, V., Bystricky, B., Beniak, J., Medvecova, L., Lagin, A., Svetlovska, D., Spanik, S., Zajac, V., Mardiak, J. and Drgona, L. (2015). Prevention of irinotecan induced diarrhea by probiotics: A randomized double blind, placebo controlled pilot study. Complementary Therapies in Medicine, 23(3): 356–362. https://doi.org/10.1016/J.CTIM.2015.03.008.

Merritt, C. R., Ong, G. T., Church, S. E., Barker, K., Danaher, P., Geiss, G., Hoang, M., Jung, J., Liang, Y., McKay-Fleisch, J., Nguyen, K., Norgaard, Z., Sorg, K., Sprague, I., Warren, C., Warren, S., Webster, P. J., Zhou, Z., Zollinger, D. R., Dunaway, D. L., Mills, G. B. and Beechem, J. M. (2020). Multiplex digital spatial profiling of proteins and RNA in fixed tissue. Nature Biotechnology, 38(5): 586–599. https://doi.org/10.1038/S41587-020-0472-9.

Mikó, E., Kovács, T., Sebő, É., Tóth, J., Csonka, T., Ujlaki, G., Sipos, A., Szabó, J., Méhes, G. and Bai, P. (2019). Microbiome-microbial metabolome-cancer cell interactions in breast cancer-familiar, but unexplored. Cells, 8(4). https://doi.org/10.3390/CELLS8040293.

Min, S., Kim, S. and Cho, S. W. (2020). Gastrointestinal tract modeling using organoids engineered with cellular and microbiota niches. Experimental & Molecular Medicine, 52(2): 227–237. https://doi.org/10.1038/s12276-020-0386-0.

Mishra, R., Rajsiglová, L., Lukáč, P., Tenti, P., Šima, P., Čaja, F. and Vannucci, L. (2021). Spontaneous and induced tumors in germ-free animals: A general review. Medicina, 57(3). https://doi.org/10.3390/MEDICINA57030260.

Mizutani, T., Yamamoto, T., Ozaki, A., Oowada, T. and Mitsuoka, T. (1984). Spontaneous polyposis in the small intestine of germ-free and conventionalized BALB/c mice. Cancer Letters, 25(1): 19–23. https://doi.org/10.1016/S0304-3835(84)80021-X.

Mo, S., Ru, H., Huang, M., Cheng, L., Mo, X. and Yan, L. (2022). Oral-intestinal microbiota in colorectal cancer: inflammation and immunosuppression. Journal of Inflammation Research, 15: 747–759. https://doi.org/10.2147/JIR.S344321.

Morrison, D. J. and Preston, T. (2016). Formation of short chain fatty acids by the gut microbiota and their impact on human metabolism. Gut Microbes, 7(3): 189–200. https://doi.org/10.1080/19490976.2015.1134082.

Murphy, E. A. and Velázquez, K. T. (2022). The role of diet and physical activity in influencing the microbiota/microbiome. In Diet, Inflammation, and Health (pp. 693–745). Academic Press. https://doi.org/10.1016/B978-0-12-822130-3.00017-X.

Nejman, D., Livyatan, I., Fuks, G., Gavert, N., Zwang, Y., Geller, L. T., Rotter-Maskowitz, A., Weiser, R., Mallel, G., Gigi, E., Meltser, A., Douglas, G. M., Kamer, I., Gopalakrishnan, V., Dadosh, T., Levin-Zaidman, S., Avnet, S., Atlan, T., Cooper, Z. A., Arora, R., Cogdill, A. P., Khan, M. A. W., Ologun, G., Bussi, Y., Weinberger, A., Lotan-Pompan, M., Golani, O., Perry, G., Rokah, M., Bahar-Shany, K., Rozeman, E.A., Blank, C. U., Ronai, A., Shaoul, R., Amit, A., Dorfman, T., Kremer, R., Cohen, Z. R., Harnof, S., Siegal, T., Yehuda-Shnaidman, E., Gal-Yam, E.N., Shapira, H., Baldini, N., Langille, M. G. I., Ben-Nun, A., Kaufman, B., Nissan, A., Golan, T., Dadiani, M., Levanon, K., Bar, J., Yust-Katz, S., Barshack, I., Peeper, D. S., Raz, D. J. Segal, E., Wargo, J. A. , Sandbank, J., Shental, N. and Straussman, R. (2020). The human tumor microbiome is composed of tumor type-specific intracellular bacteria. Science (New York, N.Y.), 368(6494): 973–980. https://doi.org/10.1126/SCIENCE.AAY9189.

Nené, N. R., Reisel, D., Leimbach, A., Franchi, D., Jones, A., Evans, I., Knapp, S., Ryan, A., Ghazali, S., Timms, J. F., Paprotka, T., Bjørge, L., Zikan, M., Cibula, D., Colombo, N. and Widschwendter, M. (2019). Association between the cervicovaginal microbiome, BRCA1 mutation status, and risk of ovarian cancer: A case-control study. The Lancet. Oncology, 20(8): 1171–1182. https://doi.org/10.1016/S1470-2045(19)30340-7.

Newman, T. M., Vitolins, M. Z. and Cook, K. L. (2019). From the table to the tumor: the role of mediterranean and western dietary patterns in shifting microbial-mediated signaling to impact breast cancer risk. Nutrients, 11(11). https://doi.org/10.3390/NU11112565.

Newsome, R. C., Yang, Y. and Jobin, C. (2022). The microbiome, gastrointestinal cancer, and immunotherapy. Journal of Gastroenterology and Hepatology, 37(2): 263. https://doi.org/10.1111/JGH.15742.

Ogunrinola, G. A., Oyewale, J. O., Oshamika, O. O. and Olasehinde, G. I. (2020). The human microbiome and its impacts on health. International Journal of Microbiology, 2020. https://doi.org/10.1155/2020/8045646.

Olsen, I. and Yamazaki, K. (2019). Can oral bacteria affect the microbiome of the gut? Journal of Oral Microbiology, 11(1). https://doi.org/10.1080/20002297.2019.1586422.

Panebianco, C., Andriulli, A. and Pazienza, V. (2018). Pharmacomicrobiomics: Exploiting the drug-microbiota interactions in anticancer therapies. Microbiome, 6(1): 92. https://doi.org/10.1186/S40168-018-0483-7/FIGURES/3.

Parahitiyawa, N. B., Jin, L. J., Leung, W. K., Yam, W. C. and Samaranayake, L. P. (2009). Microbiology of odontogenic bacteremia: Beyond endocarditis. Clinical Microbiology Reviews, 22(1): 46. https://doi.org/10.1128/CMR.00028-08.

Pereira-Marques, J., Ferreira, R. M., Pinto-Ribeiro, I. and Figueiredo, C. (2019). Helicobacter pylori infection, the gastric microbiome and gastric cancer. Advances in Experimental Medicine and Biology, 1149: 195–210. https://doi.org/10.1007/5584_2019_366.

Plummer, M., de Martel, C., Vignat, J., Ferlay, J., Bray, F. and Franceschi, S. (2016). Global burden of cancers attributable to infections in 2012: a synthetic analysis. The Lancet. Global Health, 4(9): e609–e616. https://doi.org/10.1016/S2214-109X(16)30143-7.

Poon, R. Y. C. (2016). Cell cycle control: A system of interlinking oscillators. Methods in Molecular Biology (Clifton, N.J.), 1342: 3–19. https://doi.org/10.1007/978-1-4939-2957-3_1.

Pushalkar, S., Hundeyin, M., Daley, D., Zambirinis, C. P., Kurz, E., Mishra, A., Mohan, N., Aykut, B., Usyk, M., Torres, L. E., Werba, G., Zhang, K., Guo, Y., Li, Q., Akkad, N., Lall, S., Wadowski, B., Gutierrez, J., Rossi, J. A. K., Herzog, J. W., Diskin, B., Torres-Hernandez, A., Leinwand, J., Wang, W., Taunk, P. S., Savadkar, S., Janal, M., Saxena, A., Li, X., Cohen, D., Sartor, R. B., Saxena, D. and Miller, G. (2018). The Pancreatic Cancer Microbiome Promotes Oncogenesis by Induction of Innate and Adaptive Immune Suppression. Cancer Discovery, 8(4): 403–416. https://doi.org/10.1158/2159-8290.CD-17-1134.

Rad, A. H., Abbasi, A., Kafil, H. S. and Ganbarov, K. (2020). Potential pharmaceutical and food applications of postbiotics: A review. Current Pharmaceutical Biotechnology, 21(15): 1576–1587. https://doi.org/10.2174/1389201021666200516154833.

Raven, P. H., Johnson, G. B., Mason, K. A., Losos, J. B. and Singer, S. R. (2014). How cells divide. In Biology (10th ed., pp. 187–206). McGraw-Hill.

Rea, D., Coppola, G., Palma, G., Barbieri, A., Luciano, A., Del Prete, P., Rossetti, S., Berretta, M., Facchini, G., Perdonà, S., Turco, M. C. and Arra, C. (2018). Microbiota effects on cancer: from risks to therapies. Oncotarget, 9(25): 17915–17927. https://doi.org/10.18632/ONCOTARGET.24681.

Reece, J. B., Urry, L. A., Cain, M. L., Wasserman, S. A., Minorsky, P. V. and Jackson, R. B. (2011). The cell cycle. In Campbell Biology (10th ed., pp. 232–250). Pearson.

Rezasoltani, S., Asadzadeh-Aghdaei, H., Nazemalhosseini-Mojarad, E., Dabiri, H., Ghanbari, R. and Zali, M. R. (2017). Gut microbiota, epigenetic modification and colorectal cancer. Iranian Journal of Microbiology, 9(2): 55. /pmc/articles/PMC5715278/.

Roberts, N. J., Zhang, L., Janku, F., Collins, A., Bai, R. Y., Staedtke, V., Rusk, A. W., Tung, D., Miller, M., Roix, J., Khanna, K. V., Murthy, R., Benjamin, R. S., Helgason, T., Szvalb, A. D., Bird, J. E., Roy-Chowdhuri, S., Zhang, H. H., Qiao, Y., Karim, B., McDaniel, J., Elpiner, A., Sahora, A., Lachowicz, J., Phillips, B., Turner, A., Klein, M. K., Post, G., Diaz, L. A. Jr., Riggins, G. J., Papadopoulos, N., Kinzler, K. W., Vogelstein, B., Bettegowda, C., Huso, D. L., Varterasian, M., Saha, S. and Zhou, S. (2014). Intratumoral injection of Clostridium novyi-NT spores induces antitumor responses. Science Translational Medicine, 6(249). https://doi.org/10.1126/SCITRANSLMED.3008982.

Round, J. L. and Mazmanian, S. K. (2009). The gut microbiota shapes intestinal immune responses during health and disease. Nature Reviews. Immunology, 9(5): 313–323. https://doi.org/10.1038/NRI2515.

Rous, P. (1911). A sarcoma of the fowl transmissible by an agent separable from the tumor cells. The Journal of Experimental Medicine, 13(4): 397. https://doi.org/10.1084/JEM.13.4.397.

Routy, B., Le Chatelier, E., Derosa, L., Duong, C. P. M., Alou, M. T., Daillère, R., Fluckiger, A., Messaoudene, M., Rauber, C., Roberti, M. P., Fidelle, M., Flament, C., Poirier-Colame, V., Opolon, P., Klein, C., Iribarren, K., Mondragón, L., Jacquelot, N., Qu, B., Ferrere, G., Clémenson, C., Mezquita, L., Masip, J. R., Naltet, C., Brosseau, S., Kaderbhai, C., Richard, C., Rizvi, H., Levenez, F., Galleron, N., Quinquis, B., Pons, N., Ryffel, B., Minard-Colin, V., Gonin, P., Soria, J. C., Deutsch, E., Loriot, Y., Ghiringhelli, F., Zalcman, G., Goldwasser, F., Escudier, B., Hellmann, M. D., Eggermont, A., Raoult, D., Albiges, L., Kroemer, G. and Zitvogel, L. (2018). Gut microbiome influences efficacy of PD-1-based immunotherapy against epithelial tumors. Science (New York, N.Y.), 359(6371): 91–97. https://doi.org/10.1126/SCIENCE.AAN3706.

Roy, S. and Trinchieri, G. (2017). Microbiota: a key orchestrator of cancer therapy. Nature Reviews Cancer, 17(5): 271–285. https://doi.org/10.1038/NRC.2017.13.

Rubin, H. (1955). Quantitative relations between causative virus and cell in the Rous no. 1 chicken sarcoma. Virology, 1(5): 445–473. https://doi.org/10.1016/0042-6822(55)90037-4.

Ruo, S. W., Alkayyali, T., Win, M., Tara, A., Joseph, C., Kannan, A., Srivastava, K., Ochuba, O., Sandhu, J. K., Went, T. R., Sultan, W., Kantamaneni, K. and Poudel, S. (2021). Role of gut microbiota dysbiosis in breast cancer and novel approaches in prevention, diagnosis, and treatment. Cureus, 13(8). https://doi.org/10.7759/CUREUS.17472.

Salter, S. J., Cox, M. J., Turek, E. M., Calus, S. T., Cookson, W. O., Moffatt, M. F., Turner, P., Parkhill, J., Loman, N. J. and Walker, A. W. (2014). Reagent and laboratory contamination can critically impact sequence-based microbiome analyses. BMC Biology, 12(1). https://doi.org/10.1186/S12915-014-0087-Z.

Saus, E., Iraola-Guzmán, S., Willis, J. R., Brunet-Vega, A. and Gabaldón, T. (2019). Microbiome and colorectal cancer: Roles in carcinogenesis and clinical potential. Molecular Aspects of Medicine, 69: 93. https://doi.org/10.1016/J.MAM.2019.05.001.

Savage, D. C. (1977). Microbial ecology of the gastrointestinal tract. Annual Review of Microbiology, 31: 107–133. https://doi.org/10.1146/ANNUREV.MI.31.100177.000543.

Schwabe, R. F. and Jobin, C. (2013). The microbiome and cancer. Nature Reviews Cancer, 13(11): 800–812. https://doi.org/10.1038/NRC3610.

Sears, C. L. and Garrett, W. S. (2014). Microbes, microbiota and colon cancer. Cell Host & Microbe, 15(3): 317. https://doi.org/10.1016/J.CHOM.2014.02.007.

Sears, C. L. and Pardoll, D. M. (2011). Perspective: Alpha-bugs, their microbial partners, and the link to colon cancer. The Journal of Infectious Diseases, 203(3): 306. https://doi.org/10.1093/JINFDIS/JIQ061.

Segal, L. N., Clemente, J. C., Tsay, J. C. J., Koralov, S. B., Keller, B. C., Wu, B. G., Li, Y., Shen, N., Ghedin, E., Morris, A., Diaz, P., Huang, L., Wikoff, W. R., Ubeda, C., Artacho, A., Rom, W. N., Sterman, D. H., Collman, R. G., Blaser, M. J. and Weiden, M. D. (2016). Enrichment of the lung microbiome with oral taxa is associated with lung inflammation of a Th17 phenotype. Nature Microbiology, 1(5). https://doi.org/10.1038/NMICROBIOL.2016.31.

Seidel, J. A., Otsuka, A. and Kabashima, K. (2018). Anti-PD-1 and Anti-CTLA-4 therapies in cancer: Mechanisms of action, efficacy, and limitations. Frontiers in Oncology, 8(MAR): 86. https://doi.org/10.3389/FONC.2018.00086.

Sepich-Poore, G. D., Zitvogel, L., Straussman, R., Hasty, J., Wargo, J. A. and Knight, R. (2021). The microbiome and human cancer. Science (New York, N.Y.), 371(6536). https://doi.org/10.1126/SCIENCE.ABC4552.

Sethi, V., Kurtom, S., Tarique, M., Lavania, S., Malchiodi, Z., Hellmund, L., Zhang, L., Sharma, U., Giri, B., Garg, B., Ferrantella, A., Vickers, S. M., Banerjee, S., Dawra, R., Roy, S., Ramakrishnan, S., Saluja, A. and Dudeja, V. (2018). Gut microbiota promotes tumor growth in mice by modulating immune response. Gastroenterology, 155(1): 33-37.e6. https://doi.org/10.1053/J.GASTRO.2018.04.001.

Sethi, V., Vitiello, G. A., Saxena, D., Miller, G. and Dudeja, V. (2019). The role of the microbiome in immunologic development and its implication for pancreatic cancer immunotherapy. Gastroenterology, 156(7): 2097–2115.e2. https://doi.org/10.1053/J.GASTRO.2018.12.045.

Sharma, P., Jain, T., Sethi, V., Kurtom, S., Roy, P., Tao, J., ferrantella, A., Giri, B., Edwards, D. B., Gomez, B., Ramakrishnan, S., Dawra, R., Saluja, A. and Dudeja, V. (2020). 439 cigarette smoke exposure promotes cancer progression through gut microbial dysbiosis. Gastroenterology (New York, N.Y. 1943), 158(6): S–1511. https://doi.org/10.1016/S0016-5085(20)34447-4.

Shen, J., Xiao, Z., Wu, W. K. K., Wang, M. H., To, K. F., Chen, Y., Yang, W., Li, M. S. M., Shin, V. Y., Tong, J. H., Kang, W., Zhang, L., Li, M., Wang, L., Lu, L., Chan, R. L. Y.,

Wong, S. H., Yu, J., Chan, M. T. V., Chan, F. K. L., Sung, J. J. Y., Cheng, A. S. L. and Cho, C. H. (2015). Epigenetic silencing of miR-490-3p reactivates the chromatin remodeler SMARCD1 to promote Helicobacter pylori-induced gastric carcinogenesis. Cancer Research, 75(4): 754–765.

Shi, H., Shi, Q., Grodner, B., Lenz, J. S., Zipfel, W. R., Brito, I. L. and De Vlaminck, I. (2020). Highly multiplexed spatial mapping of microbial communities. Nature, 588(7839): 676–681. https://doi.org/10.1038/S41586-020-2983-4.

Shrestha, E., White, J. R., Yu, S. H., Kulac, I., Ertunc, O., De Marzo, A. M., Yegnasubramanian, S., Mangold, L. A., Partin, A. W. and Sfanos, K. S. (2018). Profiling the urinary microbiome in men with positive versus negative biopsies for prostate cancer. The Journal of Urology, 199(1): 161. https://doi.org/10.1016/J.JURO.2017.08.001.

Sieow, B. F. L., Wun, K. S., Yong, W. P., Hwang, I. Y. and Chang, M. W. (2021). Tweak to treat: Reprograming bacteria for cancer treatment. Trends in Cancer, 7(5): 447–464. https://doi.org/10.1016/J.TRECAN.2020.11.004.

Sinha, R., Abu-Ali, G., Vogtmann, E., Fodor, A. A., Ren, B., Amir, A., Schwager, E., Crabtree, J., Ma, S., Abnet, C. C., Knight, R., White, O. and Huttenhower, C. (2017). Assessment of variation in microbial community amplicon sequencing by the Microbiome Quality Control (MBQC) project consortium. Nature Biotechnology, 35(11): 1077. https://doi.org/10.1038/NBT.3981.

Śliżewska, K., Markowiak-Kopeć, P. and Śliżewska, W. (2020). The role of probiotics in cancer prevention. Cancers, 13(1): 1–22. https://doi.org/10.3390/CANCERS13010020.

Sun, D., Chen, Y. and Fang, J. Y. (2019). Influence of the microbiota on epigenetics in colorectal cancer. National Science Review, 6(6): 1138–1148. https://doi.org/10.1093/NSR/NWY160.

Sun, J., Tang, Q., Yu, S., Xie, M., Xie, Y., Chen, G. and Chen, L. (2020). Role of the oral microbiota in cancer evolution and progression. Cancer Medicine, 9(17): 6306. https://doi.org/10.1002/CAM4.3206.

Swofford, C. A., Van Dessel, N. and Forbes, N. S. (2015). Quorum-sensing Salmonella selectively trigger protein expression within tumors. Proceedings of the National Academy of Sciences of the United States of America, 112(11): 3457–3462. https://doi.org/10.1073/PNAS.1414558112.

Takasuna, K., Hagiwara, T., Hirohashi, M., Kato, M., Nomura, M., Nagai, E., Yokoi, T. and Kamataki, T. (1996). Involvement of β-glucuronidase in intestinal microflora in the intestinal toxicity of the antitumor camptothecin derivative irinotecan hydrochloride (CPT-11) in rats. Cancer Research, 56(16): 3752–3757.

Tanoue, T., Morita, S., Plichta, D. R., Skelly, A. N., Suda, W., Sugiura, Y., Narushima, S., Vlamakis, H., Motoo, I., Sugita, K., Shiota, A., Takeshita, K., Yasuma-Mitobe, K., Riethmacher, D., Kaisho, T., Norman, J. M., Mucida, D., Suematsu, M., Yaguchi, T., Bucci, V., Inoue, T., Kawakami, Y., Olle, B., Roberts, B., Hattori, M., Xavier, R. J., Atarashi, K. and Honda, K. (2019). A defined commensal consortium elicits CD8 T cells and anti-cancer immunity. Nature 2019 565: 7741, 565(7741): 600–605. https://doi.org/10.1038/s41586-019-0878-z.

The National Cancer Institute. (2019, March 6). Outcomes & Impact of The Cancer Genome Atlas. The Cancer Genome Atlas - About the Program - NCI. https://www.cancer.gov/about-nci/organization/ccg/research/structural-genomics/tcga/history.

Thursby, E. and Juge, N. (2017). Introduction to the human gut microbiota. Biochemical Journal, 474(11): 1823. https://doi.org/10.1042/BCJ20160510.

Tian, T., Zhao, Y., Yang, Y., Wang, T., Jin, S., Guo, J. and Liu, Z. (2020). The protective role of short-chain fatty acids acting as signal molecules in chemotherapy- or radiation-induced intestinal inflammation. American Journal of Cancer Research, 10(11): 3508. /pmc/articles/PMC7716145/.

Tjalsma, H., Boleij, A., Marchesi, J. R. and Dutilh, B. E. (2012). A bacterial driver-passenger model for colorectal cancer: Beyond the usual suspects. Nature Reviews. Microbiology, 10(8): 575–582. https://doi.org/10.1038/NRMICRO2819.

Turnbaugh, P. J., Hamady, M., Yatsunenko, T., Cantarel, B. L., Duncan, A., Ley, R. E., Sogin, M. L., Jones, W. J., Roe, B. A., Affourtit, J. P., Egholm, M., Henrissat, B., Heath, A. C., Knight, R. and Gordon, J. I. (2009a). A core gut microbiome in obese and lean twins. Nature, 457(7228): 480–484. https://doi.org/10.1038/NATURE07540.

Turnbaugh, P. J., Ridaura, V. K., Faith, J. J., Rey, F. E., Knight, R. and Gordon, J. I. (2009b). The Effect of diet on the human gut microbiome: a metagenomic analysis in humanized gnotobiotic mice. Science Translational Medicine, 1(6): 6ra14. https://doi.org/10.1126/SCITRANSLMED.3000322.

Valdes, A. M., Walter, J., Segal, E. and Spector, T. D. (2018). Role of the gut microbiota in nutrition and health. BMJ (Clinical Research Ed.), 361: 36–44. https://doi.org/10.1136/BMJ.K2179.

Van Nood, E., Speelman, P., Nieuwdorp, M. and Keller, J. (2014). Fecal microbiota transplantation: facts and controversies. Current Opinion in Gastroenterology, 30(1): 34–39. https://doi.org/10.1097/MOG.0000000000000024.

Vannucci, L., Stepankova, R., Grobarova, V., Kozakova, H., Rossmann, P., Klimesova, K., Benson, V., Sima, P., Fiserova, A. and Tlaskalova-Hogenova, H. (2008). Colorectal carcinoma: Importance of colonic environment for anti-cancer response and systemic immunity. Journal of Immunotoxicology, 6(4): 217–226. https://doi.org/10.3109/15476910903334343.

Vétizou, M., Pitt, J. M., Daillère, R., Lepage, P., Waldschmitt, N., Flament, C., Rusakiewicz, S., Routy, B., Roberti, M. P., Duong, C. P. M., Poirier-Colame, V., Roux, A., Becharef, S., Formenti, S., Golden, E., Cording, S., Eberl, G., Schlitzer, A., Ginhoux, F., Mani, S., Yamazaki, T., Jacquelot, N., Enot, D. P., Bérard, M., Nigou, J., Opolon, P., Eggermont, A., Woerther, P. L., Chachaty, E., Chaput, N., Robert, C., Mateus, C., Kroemer, G., Raoult, D., Boneca, I. G., Carbonnel, F., Chamaillard, M. and Zitvogel, L. (2015). Anticancer immunotherapy by CTLA-4 blockade relies on the gut microbiota. Science (New York, N.Y.), 350(6264): 1079–1084. https://doi.org/10.1126/SCIENCE.AAD1329

Vicente-Duenãs, C., Janssen, S., Oldenburg, M., Auer, F., Lez-Herrero, I. S. G., Casado-García, A., Isidro-Hernández, M., Raboso-Gallego, J., Westhoff, P., Pandyra, A. A., Hein, D., Gössling, K. L., Alonso-López, D., De Las Arndt Rivas, J., Bhatia, S., García-Criado, F. J., García-Cenador, M. B., Weber, A. P. M., Köhrer, K., Hauer, J., Fischer, U., Sánchez-García, I. and Borkhardt, A. (2020). An intact gut microbiome protects genetically predisposed mice against leukemia. Blood, 136(18): 2003–2017. https://doi.org/10.1182/BLOOD.2019004381.

Vyas, U. and Ranganathan, N. (2012). Probiotics, prebiotics, and synbiotics: Gut and beyond. Gastroenterology Research and Practice, 2012. https://doi.org/10.1155/2012/872716.

Wang, F., Yin, Q., Chen, L. and Davis, M. M. (2018). Bifidobacterium can mitigate intestinal immunopathology in the context of CTLA-4 blockade. Proceedings of the National Academy of Sciences of the United States of America, 115(1): 157–161. https://doi.org/10.1073/PNAS.1712901115.

Wong, S. H. and Yu, J. (2019). Gut microbiota in colorectal cancer: Mechanisms of action and clinical applications. Nature Reviews. Gastroenterology & Hepatology, 16(11): 690–704. https://doi.org/10.1038/S41575-019-0209-8.

Woo, V. and Alenghat, T. (2022). Epigenetic regulation by gut microbiota. Gut Microbes, 14(1). https://doi.org/10.1080/19490976.2021.2022407.

Wu, S., Rhee, K. J., Albesiano, E., Rabizadeh, S., Wu, X., Yen, H. R., Huso, D. L., Brancati, F. L., Wick, E., McAllister, F., Housseau, F., Pardoll, D. M. and Sears, C. L. (2009). A human colonic commensal promotes colon tumorigenesis via activation of T helper type 17 T cell responses. Nature Medicine, 15(9): 1016–1022. https://doi.org/10.1038/NM.2015.

Yang, J., Tan, Q., Fu, Q., Zhou, Y., Hu, Y., Tang, S., Zhou, Y., Zhang, J., Qiu, J. and Lv, Q. (2017). Gastrointestinal microbiome and breast cancer: correlations, mechanisms and potential clinical implications. Breast Cancer (Tokyo, Japan), 24(2): 220–228. https://doi.org/10.1007/S12282-016-0734-Z.

Yang, Q., Liang, Q., Balakrishnan, B., Belobrajdic, D. P., Feng, Q. J. and Zhang, W. (2020). Role of dietary nutrients in the modulation of gut microbiota: A narrative review. Nutrients, 12(2). https://doi.org/10.3390/NU12020381.

Ying, M., Yu, Q., Zheng, B., Wang, H., Wang, J., Chen, S., Nie, S. and Xie, M. (2020). Cultured Cordyceps sinensis polysaccharides modulate intestinal mucosal immunity and gut microbiota in cyclophosphamide-treated mice. Carbohydrate Polymers, 235. https://doi.org/10.1016/J.CARBPOL.2020.115957.

Yu, L. X., Yan, H. X., Liu, Q., Yang, W., Wu, H. P., Dong, W., Tang, L., Lin, Y., He, Y. Q., Zou, S. S., Wang, C., Zhang, H. L., Cao, G. W., Wu, M. C. and Wang, H. Y. (2010). Endotoxin accumulation prevents carcinogen-induced apoptosis and promotes liver tumorigenesis in rodents. Hepatology (Baltimore, Md.), 52(4): 1322–1333. https://doi.org/10.1002/HEP.23845.

Zengul, A. G., Demark-Wahnefried, W., Barnes, S., Morrow, C. D., Bertrand, B., Berryhill, T. F. and Frugé, A. D. (2021). Associations between dietary fiber, the fecal microbiota and estrogen metabolism in postmenopausal women with breast cancer. Nutrition and Cancer, 73(7): 1108–1117. https://doi.org/10.1080/01635581.2020.1784444.

Zerdan, M. B., Niforatos, S., Nasr, S., Nasr, D., Ombada, M., John, S., Dutta, D. and Lim, S. H. (2022). Fecal microbiota transplant for hematologic and oncologic diseases: Principle and practice. Cancers, 14(3). https://doi.org/10.3390/CANCERS14030691.

Zhan, Y., Chen, P. J., Sadler, W. D., Wang, F., Poe, S., Núñez, G., Eaton, K. A. and Chen, G. Y. (2013). Gut microbiota protects against gastrointestinal tumorigenesis caused by epithelial injury. Cancer Research, 73(24): 7199–7210. https://doi.org/10.1158/0008-5472.CAN-13-0827.

Zhang, X., Li, C., Cao, W. and Zhang, Z. (2021). Alterations of gastric microbiota in gastric cancer and precancerous stages. Frontiers in Cellular and Infection Microbiology, 11: 69. https://doi.org/10.3389/FCIMB.2021.559148/BIBTEX.

Zheng, J. H., Nguyen, V. H., Jiang, S. N., Park, S. H., Tan, W., Hong, S. H., Shin, M. G., Chung, I. J., Hong, Y., Bom, H. S., Choy, H. E., Lee, S. E., Rhee, J. H. and Min, J. J. (2017). Two-step enhanced cancer immunotherapy with engineered Salmonella typhimurium secreting heterologous flagellin. Science Translational Medicine, 9(376). https://doi.org/10.1126/SCITRANSLMED.AAK9537.

Chapter 7

Microbiomes and Probiotics

Firasat Hussain,[1] *Shafeeq Ur Rehman,*[1,#] *Muhammad Naveed Nawaz,*[1,#] *Kashif Rahim,*[1] *Ahmed Abdelmoneim,*[2] *Kamal Niaz,*[2,3,*] *Murad Muhammad*[4] and *Wen-Jun Li*[4,5,*]

Introduction

Viruses, bacteria, archaea, fungi, and protozoa comprise the human microbiome. In terms of the number of microbial cells and DNA, the bulk of the human microbiome is composed of bacterial species related to humans. Bacterial sequences account for over 99% of traced DNA sequencing in healthy people, as depicted in a recent study (Versalovic 2013). *Firmicutes, Proteobacteria, Actinobacteria* and *Bacteroidetes* are the primary phyla in human-associated bacterial ecosystems. Various smaller phyla are also present (Zhernakova et al. 2016). A recent investigation

[1] Department of Microbiology, Faculty of Veterinary Science, Cholistan University of Veterinary & Animal Sciences, Bahawalpur-63100, Pakistan.

[2] Department of Comparative Biomedical Sciences, Louisiana Animal Disease Diagnostic Laboratory, School of Veterinary Medicine, Louisiana State University, Baton Rouge, LA 70803, USA.

[3] Department of Pharmacology & Toxicology, Faculty of Bio-Sciences, Cholistan University of Veterinary & Animal Sciences, Bahawalpur-63100, Pakistan.

[4] State Key Laboratory of Desert and Oasis Ecology, Xinjiang Institute of Ecology and Geography, Chinese Academy of Sciences, Urumqi 830011, PR China.

[5] State Key Laboratory of Biocontrol and Guangdong Provincial Key Laboratory of Plant Resources, School of Life Sciences, Sun Yat-Sen University, Guangzhou 510275, PR China.

* Corresponding authors: liwenjun3@mail.sysu.edu.cn; kniaz@lsu.edu, kamalniaz1989@gmail.com

# Authors contributed equally in the drafting of this manuscript.

into the biogeography of the human microbiome found that newborns and adults harbour 30 phyla in 18 different bodily locales (Zhou et al. 2013). Studies of cultured and uncultured bacteria estimate that more than 50 bacterial phyla are connected with humans (Pace 2009). Changes in function may correspond to changes in physiology at every bodily part because a segment or habitat consists of a specific microbiome and microbiological makeup.

Human eating habits and nutritional components unquestionably affect the bacteria makeup and, probably, function in the colon. In Western African children fed a high-fibre, plant-based diet, bacterial species such as *Xylanibacter* and *Prevotella* were found quite commonin their stools. These bacteria may help in the digestion of plant parts and dietary fibres. Southern European children consuming a western diet seldom or never encounter these two genera, even though genera have genomics and metabolic ways that can degrade dietary xylan and cellulose (De Filippo et al. 2010). A recent study also described significant variations between children's fecal microbiomes in Bangladesh and the United States (Lin et al. 2013). Basic long-term dietary variations appear to alter gut microbial makeup, even while major temporary dietary alterations cannot significantly affect gastrointestinal microbiota (Wu et al. 2011). In addition to food and probiotic usage, prebiotics or symbiotic mixtures may cause a change in the composition of the intestinal microbiota (Gibson et al. 1995). Gut microorganisms create and convert a wide range of chemicals that impact a person's physiology, immunology, and perhaps susceptibility or resistance to disease. Vitamin B12 production is an example of a biosynthesis pathway in the gut microbiota and probiotics – cobalamin (Santos et al. 2008) and thiamine (Saulnier et al. 2011). For instance, transformation in luminous plant lignins may be transformed into enterolignins, and vitamin K1 (phylloquinone) can be transformed into vitamin K2 and its analogs (menaquinones) (Dairi 2012). Biogenic amines are produced through acid decarboxylation processes, such as glutamate/GABA and histidine/histamine (De Biase and Pennacchietti 2012, Thomas et al. 2012), and tiny fatty acid chain formation from dietary fibreand carbs (van Zanten et al. 2012).

Probiotics are alive microorganisms administered adequately and impose health benefits on the host (Hill et al. 2014). The microbes, predominantly bacteria (including yeast strains), are present during fermentation usually and can also be bought as dietary supplements. They can also be added to other food products. Prebiotics and probiotics must not be confused with each other; prebiotics are often polysaccharides like oligosaccharides and inulin, which intestinal bacteria utilize as food (Gibson et al. 2017). Commercialized substances that include prebiotics and probiotics are referred to as synbiotics. Probiotics are also excluded from substances made from or containing dead microbes consisting of nucleotides, polysaccharides, proteins and polypeptides. The unique strain of a probiotic is its unique identification and includes the genus, species, sub-species and Unicode strain-description. The seven principal genera of bacteria most commonly used commercially are *Enterococcus, Bifidobacterium, Saccharomyces, Lactobacillus, Bacillus* and *Streptococcus*. The nomenclature of the commercial probiotic strains is shownin Table 1 (Merenstein and Salminen 2017).

Table 1. The Unicode strain of probiotic microorganisms.

Genus	Species	Strain	Designation
Lactobacillus	*rhamnosus*	*LGG*	*GG*
Bifidobacterium	*animalis*	*Bifidus regularis*	*DN-173 010*

Mechanism of Action

Various microbes (comprising bacteria, fungi, viruses, protozoa and archaea) flourish within the human gastrointestinal system. These microorganisms, which comprise the intestinal microflora as a whole, can affect health and illness in people. Probiotics often work in the GI tract, where they may impact the gut flora. Beneficial microbes can fill the intestinal lumen in unusual ways based on the probiotic strain, surrounding microflora, and GIT location (Zmora et al. 2018). Probiotics also provide various general, species-specific and strain-specific health benefits. Strains, species and even taxa of probiotic supplements differ significantly in non-specific processes.

These processes include the synthesis of beneficial metabolites such as minute fatty acid chains, lowering of pH in the large intestine, and suppression of pathogenic microbial development in the digestive system (by facilitating invasion resistance, enhancing intestinal transit, or aiding in the normalization of a peeved microbiota). Some species-specific systems include the production of vitamins, enzymatic activity, toxin mitigation, bile salt regulation, and reigning of the intestinal barrier. Examples of processes specific to strains involve immune-alteration, cytokine regulation, and those affecting the endocrine system. These processes are unusual and employed by just a few variations of a particular species (Guarner et al. 2017). According to research, probiotics alter the gut environment by triggering unimmunized responses to prospective invaders. The World Gastroenterology Organization (WGO) lists some probable probiotic mechanisms of action as follows:

- Mobilize nearby macrophages
- Alter the profiles of cytokines

- Make people less susceptible to dietary antigens
- Food digestion and nutrient competition
- Change the pH in the area
- Manufacture bacteriocins
- Neutralize superoxide radicals
- Encourage the synthesis of epithelial mucin
- Raise intestinal barrier integrity
- Compete to stick
- Alter pathogen-produced poisons (Guarner et al. 2012)

Sources of Probiotics

Various live microbial cultures multiply and carry out metabolic processes to create fermented foods. Many of these meals are abundant in living and potentially helpful microorganisms. Some fermented foods, such as most commercial pickles and rye bread, are treated following fermentation and are not comprised of active microbes that are fit for human consumption. Probiotic microorganisms like *Lactobacillus bulgaricus* and *Streptococcus thermophilus* are commonly found in commercial curd, another fermented food.

Many fermented foods, like yogurt, include living microbes that typically remain viable throughout the product's shelf life. They usually do not make it past the stomach and might be unable to endure the destruction of bile salts and hydrolytic enzymes in the small intestine, which might keep them from getting to the distal gut. However, probiotic strains found in yogurt and other foods survive the intestinal transit. Pickles, sore apple cider vinegar, kimchi, sauerkraut, miso, kombucha and many kinds of cheese are fermented foods but do not typically contain verified probiotic microorganisms (Kailasapathy and Chin 2000). Additionally, several probiotic strains and dosages are available as dietary supplements in the form of pills, pellets, fluids, etc. Instead of just one strain, these goods frequently contain mixed cultures of live microorganisms. Because many commercial items that include 'probiotics' have their effects unexplored in investigations, it has become difficult for consumers ignorant of experimental studies to find data-supported commercial products. Nevertheless, several institutes have thoroughly examined the present situation and provided suggestions on certain beneficial microbes –the best brand, formulation and dosage for treating a range of sicknesses (Depoorter and Vandenplas 2022).

Probiotics and their Significance

Probiotics may provide beneficial health effects, which is the subject of several scientific studies. The seven health issues that are the focus of this study are infectious diarrhea, inflammatory bowel disease (IBD), atopic dermatitis, atopic dermatitis, antibiotic-induced diarrhea, obesity, hypercholesterolemia and irritable bowel syndrome.

Atopic Dermatitis

Atopic dermatitis, one of the most prevalent chronic dermatological conditions, affects 1% to 3% of grown-ups and 15% to 20% of children globally; it is among the most common eczema (Avena-Woods 2017). Numerous meta-analyses have revealed the results of microbial research that assessed the benefits of various types and strains of bacteria on the reduction of atopic dermatitis. According to these investigations, probiotic revelation in the course of gestation and the earlier life of a child may decrease the onset of atopic dermatitis. For instance, a study comprised 6907 children and newborns who were given probiotics for 14 days to 7 months during pregnancy prenatally and for 2–13 months orally following birth in 2018 (Li et al. 2019a). Probiotic therapy with individual strains or mixtures of *Propionibacterium, Lactobacillus* and *Bifidobacterium* strains between 6 months and 9 years dramatically decreased the risk of dermatitis from 34.7% in the placebo group to 28.5% in the probiotic group.

According to subgroup analysis, probiotics given both during pregnancy and after birth significantly decrease the incidence of dermatitis. However, probiotics taken solely during pregnancy or after birth do not. Depending on the probiotic strain, the probiotic treatment produces varying benefits. For instance, supplementing with either *Lactobacillus paracasei* or *Lactobacillus rhamnosus* dramatically decreases the occurrence of eczema, while no reduction is shown by using *Lactobacillus acidophilus* or *Lactobacillus reuteri*. Contrarily, a study of five randomized clinical tests totalling 889 individuals depicted that no matter how the supplementation regimen was scheduled, the supplementation of *LLG* (*Lactobacillus rhamnosus GG*) did not decrease atopic dermatitis in children under four (Szajewska and Horvath 2018).

According to most documented meta-analyses, probiotics marginally reduce the symptoms and appearance of atopy in babies and children. For example, probiotic medication for up to eight weeks dramatically decreases scoring atopic dermatitis (SCORAD) levels in patients with atopy, indicating lower symptom intensity, according to a recent study of 13 Randomized Clinical Trials (RCTs) with 1,070 participants who were

18 years of age or younger (Huang et al. 2017). According to subgroup analysis, probiotics show protective advantages in children aged 1 to 18 (nine studies) but not in neonates under 1 year (five trials). Additionally, probiotic therapy using the probiotic strains *Lactobacillus, Lactobacillus fermentum,* or a combination of probiotic strains significantly reduces the SCORAD levels in kids. However, *Lactobacillus rhamnosus GG* and *Lactobacillus plantarum* have no effect (Zhao et al. 2018).

Even though these treatments can offer only a little comfort from an ailment, research generally points to the possibility that probiotic usage can lower the likelihood of contracting atopic dermatitis and result in considerable drops in atopic dermatitis levels. Probiotic outcomes change depending on the type applied when they are provided, and the patient's age, making medical advice challenging in these circumstances.

Pediatric Acute Infectious Diarrhea

A rise in gastrointestinal urgency and loose or watery feces are two characteristics commonly associated with acute diarrhea. Additionally, acute diarrhea, which usually lasts no longer than seven days, may sometimes be accompanied byvomiting or fever (Guarino et al. 2014). Single and multi-strain probiotics significantly lower the period of highly transmissible diarrhea by roughly 25 hours, according to a controlled study of 63 RCTs with 8,014 subjects. They also drop the likelihood that diarrhea might persist for over four days by 59%, and patients who receive probiotics have around one less digestion on day 2 than those who do not (Allen et al. 2010). With a daily dosage of at least 1010 CFU, infectious diarrhea is best treated with *Lactobacillus rhamnosus GG,* according to a review of randomized clinical trials that were randomly chosen and involved 2,444 individuals (Caffarelli et al. 2015). A review of 22 randomized clinical studies with over 2,000 individuals aged from a month to 15 years showed that *Saccharomyces boulardii*decreases the diarrheal onset (Feizizadeh et al. 2014). Research also suggests that *Saccharomyces boulardii* and *LGG* reduce the infectious diarrheal period by a day (Freedman et al. 2018). This is predicated on the idea that to show the efficacy of an intervention, there must be at least two sufficient and carefully conducted studies, each of which is believable independently. According to recent studies, probiotics may not be effective in developed nation emergency rooms, as most bouts of acute infectious diarrhea are self-limiting and do not need any more care beyond rehydration therapy. Therefore, there is disagreement over whether using probiotic supplements to treat acute viral diarrhea is cost-effective (Allen et al. 2010, Freedman et al. 2018, Schnadower et al. 2018).

Antibiotic-associated Diarrhea

Antibiotic use is another common factor in acute-onset diarrhea. The intestinal microflora is commonly changed by antibiotic therapy, and by diminishing microbial variety, it is possible to lose microbial metabolism and colonization resistance and enhance gastrointestinal motion (Cha et al. 2012). Approximately one-third of persons who take antibiotics get diarrhea due to antimicrobials (Silverman et al. 2017). People who get inpatient care have a far higher chance of acquiring AAD (antibiotic-associated diarrhea) than people who receive outpatient care. In the same vein, babies less than 2 and adults over 65 years of age are more vulnerable to acquiring AAD than other kids and adults. Penicillin and erythromycin, for example, are two medications that are more frequently linked to AAD than others (Cha et al. 2012, Silverman et al. 2017). According to meta-analyses, taking any probiotic species and strains may reduce antimicrobial diarrhea incidence by 51% (Blaabjerg et al. 2017).

However, the advantages of utilizing probiotics to avoid diarrhea vary depending on the category of antimicrobials that brought the illness, probiotic strains applied, the user's age (child, young adult, older adult), and if they are receiving inpatient or outpatient therapy. In children and adults between 18 and 64 age, probiotic usage has been linked to the reduction of AAD incidence, but not in people aged 65 or over(Jafarnejad et al. 2016). *Saccharomyces boulardii* and *Lactobacillus rhamnosus GG* both result in a decrease in diarrheal incidence. In a comprehensive study of 12 randomized clinical trials involving more than 1400 subjects, therapy by *LGG* for 10–90 days lowered the incidence of antibiotic-associated diarrhea from 22.4% to 12.3% compared to the control (Szajewska and Kołodziej 2015). Only the differential in children was clinically significant, as shown by separate examinations of 445 children and 1,052 adults. The incidence of AAD in children was considerably reduced by 71% by taking 1 to 2 x 1010 CFU/day of *Lactobacillus rhamnosus GG*, even though the optimal dosage was uncertain (Szajewska and Kołodziej 2015). Probiotics should be started within two days of the initial antibiotic dose for best results (Szajewska and Kołodziej 2015).

Role of Probiotics in Gut Microbiota Homeostasis

Comparing microbes in slime or adhering to the gut lining, characterized as the parietal microflora with microorganisms residing in moving food and excrement, is called luminal microbial community. It becomes clear how the microbiota arises as a targeted habitat and changes from area to area (Caballero et al. 2015, Sonnenburg et al. 2006). Depending upon the effect of nutrition, exposure to ingested probiotic bacteria, gut environmental circumstances, and other host-related variables, the microbiota composition is dynamic and personalized and may 'temporarily' incorporate some new strains in the ecosystem (Derrien and van Hylckama Vlieg 2015, Zhang et al. 2015). Taking probiotics to boost immunity and overall healthhas often been suggested. Diabetes and obesity are examples of metabolic illnesses that may be influenced by the intestinal microbiota (Choi et al. 2017).

Research has shown that probiotics improve the prognosis and treatment of such illnesses (Qin et al. 2010). Moreover, the genetic and microbiological traits of specific probiotic strains are significant. More research is required to fully comprehend how the consumption of these probiotics affects the human gut microbiome. The concept behind probiotics, how they function, and how they relate to conditions linked to the gut microbiota are briefly summarized in this chapter (Wang et al. 2019). By improving the balance of the gut flora, probiotics are widely used to maintain thewell-being of the human gut (Sanders et al. 2011). The microbiome is balanced by decreasing the number of dangerous bacteria that cannot live in an acidic condition while increasing the number of helpful bacteria that thrive in an acidic condition. Figure 1 illustrates the connection between sickness, microbiome and probiotics.

Probiotics may have favourable immune-modulating effects and nutritional benefits in individuals with inflammatory bowel illness. Probiotics may have beneficial effects leading to therapeutic advantages, based on a recent investigation into the anti-allergic benefits of probiotics in individuals suffering from IBD (Bae et al. 2018, Cremonini et al. 2002, Lorea Baroja et al. 2007). Intestinal health in individuals is correlated with the variety and abundance of microbes that make up the human gut microbiota (Clemente et al. 2012). The intestinal flora consists of approximately 100 trillion microbes and carries out several metabolic functions for the host (Ley et al. 2006). For instance, the human host cannot digest complex carbohydrates received from plants because it lacks the required enzymes for breaking down carbohydrates contained in plant material. Hence, the gut microbiota is essential for their breakdown (Flint et al. 2012). Most anaerobic bacteria in the large intestine might be harmful or beneficial (Apajalahti 2005).

Additionally, the notion has been established that the synthesis and breakdown of short-chain fatty acids (SCFAs), like butyrate, propionate and acetate, control the makeup of the gut microbiome. Furthermore, it is understood that gut flora aids in the growth of the intestines, maintains homeostasis, and protects the body from dangerous microbes. According to studies, ulcerative colitis (UC), colorectal cancer, AAD, IBD, Crohn's disease and metabolic illnesses (including obesity and diabetes) have

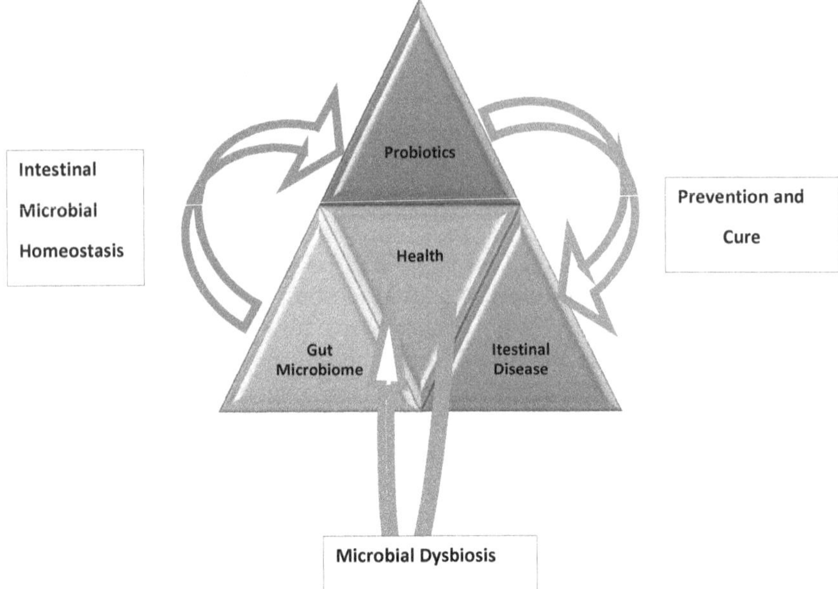

Figure 1. The use of probiotics for treating intestinal dysbiosis in the intestinal microbiota is substantial. Intestinal dysbiosis between host-beneficial and host-harmful bacteria has been recognized to be highly linked with several gastrointestinal disorders, including AAD, IBD, Crohn's disease (CD) and CRC. Inflammation, diarrhea, and even colon cancer can occur when the delicate equilibrium of the intestinal microbiota is upset. It has been suggested that administering probiotics can alleviate several digestive diseases. Since probiotics are known to have antibacterial, anti-inflammatory, and even anti-carcinogenic characteristics, they may aid in reestablishing the gut microbiota's usual balance. The symptoms of certain intestinal illnesses may also be diminished or reduced by this recovery. The relation among probiotics, the gut microflora, and gastrointestinal health and illness is now widely acknowledged due to this understanding (Lorea Baroja et al. 2007).

been proven to be influenced by gut microbiota dysbiosis (Qin et al. 2010, Wen et al. 2018).

Human Gut Microbiota Dysbiosis

An abnormality in the gut flora's composition and operation is called microbiological dysbiosis (Carding et al. 2015). Bacterial infections, dietary changes, and drugs contribute to this pervasive problem in the contemporary era. Recent research has demonstrated a link between intestinal diseases (such as celiac disease, IBS and IBD) and intestinal dysbiosis of microbes. Beneficial bacteria in the digestive system prevent dangerous bacteria from invading and multiplying by competing for resources and space (Ouwehand et al. 1999). It is interesting to note that following antibiotic therapy, which causes gut microbiota dysbiosis, probiotics are essential for restoring the human intestine's microbial balance and shielding patients against infections.

Probiotics and the Gut Barrier Function

Even though gut tissues consistently face their antigens and metabolites, we have an excellent symbiotic connection with these bacteria. This arrangement is made feasible by several different factors. Under normal conditions, the gut barrier performs incredibly effectively in complex, multidimensional processes that include antibiotics, immunoglobulin A, tight junction peptides, and the presence of a mucus layer (Riiser 2015, Wells et al. 2017). There is a link between intestinal microbes and the immune system that microbial components or receptors also regulate energy, glucose and lipid metabolism, in addition to the more typical immunological functions (Duparc et al. 2017, Everard et al. 2014, Yang et al. 2007). An elevated amount of circulating lipopolysaccharides (LPS) has been seen in models of both genetically transmitted obesity and diabetes (Yang et al. 2007). Moreover, a little increase in the blood levels of LPS has proved to be a crucial element aiding the start of a minor inflammation that ultimately results in insulin resistance in case of obesity and similar abnormalities. Additional substantial human cohorts later confirmed this result (Amar et al. 2008, Gummesson et al. 2011, Lassenius et al. 2011).

Furthermore, a link between the development of prediabetes and intestinal microflora has not yet been shown with respect to metabolic endotoxemia. However, alterations in the makeup and activity of microorganisms are strongly linked to the primary mechanism of the onset of metabolic endotoxemia (Cani et al. 2008, Dewulf et al. 2013). It has been asserted that further to the distinctive modifications in the makeup of the human microbiome, T lymphocytes increase in the lumen of fatty persons who consume high-lipid meals, a fact that is associated with illness (Magalhaes et al. 2015, Monteiro-Sepulveda et al. 2015). Along similar lines, several studies have demonstrated that fecal material transplantation and its application can change the intestinal flora, which may impact the host's metabolism (Khan et al. 2014, Vrieze et al. 2012). A similar study primarily targets many strains – newly discovered probiotics such as *Faecalibacterium prausnitzii* and *A. munciphila* thought to be next-level probiotics (O'Toole et al. 2017).

Saccharomyces boulardii is the subject of the substantial investigation into inflammatory gut diseases and gut barrier failure. Additionally, the therapeutic benefits of this yeast are explained by its nourishing effects on the gut mucosa and its antibacterial and antitoxin properties (Szajewska et al. 2010). More recently, it has been demonstrated that *A. muciniphila* functions as a barrier, restoring tight junction protein and tight junction stiffness and producing specific antibacterial lipids (Brookes et al. 2011, Plovier et al. 2017). These behaviours show that *A. muciniphila* interferes with the gut defence of the host by interacting with several parts and is still regarded as next-level probiotics (Cani and de Vos 2017). People have received *A. muciniphila* for the first time in certain studies. Obese adults administered with *A. muciniphila* for 90 days showed lower LPS of plasma, higher sensitivity to insulin, and lower ratios of hepatic and systemic allergic markers, according to proof-of-concept research. As a result, despite not being a probiotic, *A. muciniphila* is shown to encourage early results in people (Nikonovas et al. 2020). *F. prausnitzii* is an anti-inflammatory commensal bacterium identified in gut microbiota. It is low in obese people, those with type 2 diabetes, and those with IBD (Sokol et al. 2008).

How do Probiotics Alter the Intestinal Microbiota?

The microbiome is a concentrated population that differs locally. Interestingly, when bacteria harbour in the intestine or the mucus layer, it is called a parietal microbiome; germs dwelling in diet and feces constitute the luminal microbiome (Caballero et al. 2015, Lee et al. 2013, Sonnenburg et al. 2006). Depending upon the effect of nutrition, exposure to ingested probiotic bacteria, gut environmental circumstances, and other host-related variables, the microbiota composition is dynamic and personalized and may 'temporarily' add a few new strains to the ecology (Derrien and van Hylckama 2015, Zhang et al. 2016). The luminal microbiota changes following probiotic therapy, indicating endurance throughout the passage of the gastrointestinal tract. *Lactobacillus* and *Bifidobacterium* strains are found in the feces of more than 90% of the subjects (Larsen et al. 2011, Rutten et al. 2015).

In a separate investigation, *L.rhamnosus* DR20 controlled 3.105 CFU of *lactobacilli* per gram of feces in 6/10 participants. However, only 10% of the individuals remained colonized 2 months after the treatment ended (Tannock et al. 2000). After receiving *Lactobacilli* therapy, microflora is discovered in patients' feces as stool samples subsequently include more different bacteria (Alander et al. 1997). Do the bacteria in luminal microbiota have any therapeutic significance? It demonstrates that *Enterococcus* strains are recovered more frequently after receiving *L. rhamnosus* therapy (Tannock et al. 2000). These changes may alter the targeted population of the harboring microflora. A higher concentration of potentially butyrate-producing bacteria is found after consuming yogurt, which includes Bifidobacterium species, which impacts systemic metabolism. An increase in enzymatic pathways linked to glucose metabolism is also linked to probiotic-treated fermented milk (Cha et al. 2012, Johansson et al. 1998). Systemic metabolism, including insulin resistance, may be substantially affected by probiotic changes to the parietal microbiota. Probiotic has beneficial bacteria that are vital in treating chronic inflammatory diseases, especially Crohn's disease (Alander et al. 1997).

Probiotics are a group of helpful bacteria that also significantly influence the regulation of the gut flora (RodrÍGuez et al. 2011). Mice fed high-fat diets have specific changes in their gastrointestinal microflora, which can be connected to several illnesses. When probiotics are injected into obese animals, it has been observed that the gut flora's *Firmicutes*are reduced and its *Actinobacteria* increase. Mice with gut flora dysbiosis can be treated with probiotics to reduce inflammation. Intestinal flora is an essential factor in managing inflammatory illnesses (Bagarolli et al. 2017). We can assume that the control of probiotics over the gut flora can successfully cure associated inflammatory illnesses. Researchers have found higher levels of short-chain fatty acids in the gastrointestinal microbiota of mice treated with probiotics, such as *Oscillibacter* and *Prevotella,* compared to the control group (Li et al. 2016). Due to their significance as crucial metabolites of gut bacteria, SCFAs have tremendous scientific relevance. Short-chain fatty acids can enhance the immune system and work as a chemical messenger for the gut-brain axis (Guevara-Romero et al. 2022). Probiotics can increase the number of SCFA-producing bacteria in the gut. By doing this, we might encourage a healthy lifestyle and fortify our immune system. Probiotics also undeniably regulate gut flora, as demonstrated by many investigations. Probiotics, for instance, may counteract the loss of gut flora variety by antibiotic usage (Oh et al. 2022). Another study found that yogurt improved the intestinal function of people with inflammatory bowel disease (IBD) by increasing *Lactobacilli* and *Bifidobacterium* species (Shadnoush et al. 2015).

Inhibition of Pathogens by Probiotics

Numerous bacteria live in the gastrointestinal tract of humans, including both beneficial and detrimental bacteria. Pathogenic microbes may change the gut microbiota's homeostasis, increasing the likelihood of contracting associated disorders (Iqbal et al. 2021). In earlier research, probiotics have been shown to protect the digestive tract by suppressing harmful microorganisms. As a result, the scientific community has been very interested in how probiotics influence gut pathogens and the mechanisms that underlie them. Probiotics suppress pathogenic bacteria in several ways, such as by promoting the function of the epithelial barrier, producing antimicrobial compounds, and blocking pathogens from harbouring and masking receptor sites by competition (Nair et al. 2017).

On the other hand, a significant advantage of probiotics is the competitive exclusion of infections (Khaneghah et al. 2020). The beneficial strain of *E. coli* suppresses *enterohemorrhagic E. coli* (EHEC) by secreting DegP, a bifunctional periplasmic protein (Fang et al. 2018). Through extracellular DegP activity, the probiotic *Escherichia coli* outperforms the colonies of pathogens during the colonization of different bacterial species. In a different experiment, probiotics were shown to release antimicrobial compounds that interfered with *Helicobacterpylori's* ability to adhere to epithelial cells by creating competition adhesion sites and nutrients (Qureshi et al. 2019). On the other hand, probiotics are also essential for releasing antimicrobial agents. Probiotics can emit different kinds of organic fatty acids during the fermentation of carbohydrates. The primary antimicrobial substances thought to be in charge of their inhibition effect against infections have been identified as organic acids. Because of the pH decrease, fatty acids act as antimicrobial agents (Khaneghah et al. 2020). Interestingly, some studies have also discovered a way to eliminate intestinal pathogens by inhibiting the pathogen signal system.

The Human Microflora Microbiota

The collection of microorganisms discovered in the extracellular matrix and body cells, as well as their associated morphological positions, is referred to as the 'human microbiome' (Marchesi and Ravel 2015), such as the gastrointestinal tract, lungs, epidermis, glandular organs, semen, womb, Graafian follicle, sputum, consensual membrane and pterygium. The word 'human microbiome' is the aggregate of genes harbouring bacteria, even though 'human metagenome' is similar to 'human microbiome' (Willey et al. 2011). Humans are home to various bacteria (Sender et al. 2016). Some microbes do not cause any harmful effect on us but merely take shelter in our bodies. Other microorganisms coexist with us mutually (Willey et al. 2011). Our body oxidizes the metabolites certain non-pathogenic microorganisms produce through the action of FMO3, which can affect human hosts in turn (Falony et al. 2015). Although the functions of certain microbes are recognized to be advantageous to the human host, most of them are still poorly understood. As expected, these organisms reside in our bodies and do not cause illness. They are also referred to as normal flora or normal microbiota.

Bacteria

Microorganisms like bacteria and yeast have colonies on the epidermis and the mucous layer in different body regions. They act as an essential

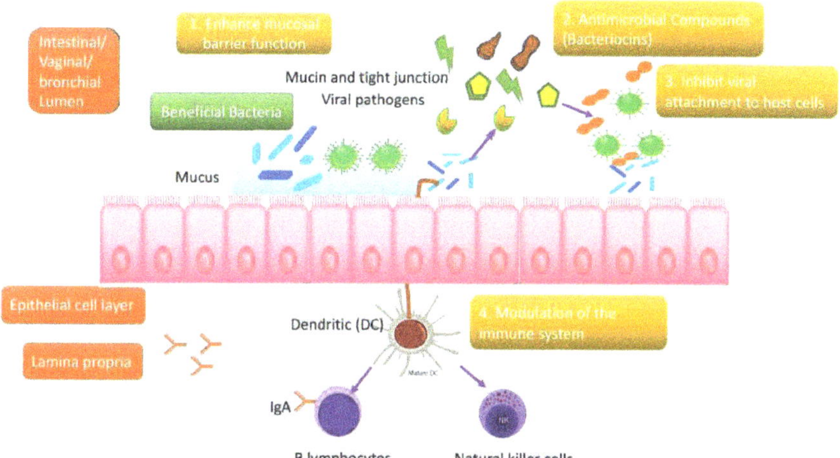

Figure 2. Management of microbiota by mucous membranes. 1. Wet epithelial surfaces, such as those in the GI tract, vagina and lungs, are coated with a mucous layer that contains glycoproteins known as mucins, which operate as a physical barrier between invasive pathogenic microbes and host epithelial cells. The GI microbiota impacts mucin formation and seem to have antiviral characteristics. Several lactic acid bacteria have been shown to control tight junctions and maintain typical mucosal permeability. 2. Certain bacteria, including Bacillus species, create virucidal antimicrobial substances (bacteriocins) called subtilosin and other similar substances. 3. Various techniques can be used to prevent the attachment of viruses to the host epithelial cells. 4. Immune system modification: Probiotics encourage the production of interferon-alpha (IFN-a) by plasmacytoid dendritic cells, triggering the cytotoxic activity of natural killer (NK) cells (innate immune system), which are crucial in viral infections. Moreover, probiotic bacterial strains have been demonstrated to increase the synthesis of antiviral immunoglobulins (made by B lymphocytes - the adaptive immune system) by an unidentified method, presumably by stimulating intestinal epithelial cells or immunocytes.

component for good human health. Diseases may occur if microbes take shelter beyond the limits or colonise the body's unexpected parts due to poor cleanliness and bruising, leading to bacterial infection. Each human carries hundreds of different bacterial species, and each body part has its distinct ecosystem, according to the human microbiome project (Figure 2). The epidermis and vaginal regions show less variety than the mouth and stomach, which exhibit the most variety. Each person has a definite mixture of microbes in quantity and quality at every location in their body. The oral cavity nurtures the second largest and most diverse microbiota, including bacteria, fungi, viruses and protozoa. Although there are thousands of definite species of microbes that live in the stomach, including a few phyla, *Bacteroidota* and *Bacillota* prevail despite the presence of *Verrucomicrobiot, Pseudomonodota, Actinobacteriota, Cyanobacterium* and *Fusobacteriota* (Sommer and Bäckhed 2013).

Archaea

They are present in the stomach in smaller populations than bacterial species. The most common category is methanogens, especially *Methanobrevibacter smithii* and *Methanosphaera stadtmanae* (Duncan et al. 2007). Furthermore, there are varieties of methanogen colonies, and half of the population has distinct ones that can be seen easily (Florin et al. 2000). However, there has been speculation linking the presence of several methanogens to human periodontal disease. As of 2007, no confirmed archaeal infections have been found (Lepp et al. 2004).

Fungi

Particularly, fungi known as yeasts are sheltered in the gastrointestinal tract of humans. The most researched species of fungus is *Candida*, because of its ability to show pathogenesis in immune-deficient patients. Some yeasts are found on the skin, feeding on oil production (Cui et al. 2013).

Viruses

Viruses, particularly bacterial viruses, invade many body regions (bacteriophages). The epidermis, intestines, lungs and oral cavity are only a fewcolonized regions (Hannigan et al. 2015). Numerous illnesses have been connected to viral communities, which do not merely resemble other microbial communities (Monaco et al. 2016).

Body Parts

Epidermis

The four phyla – *Bacilotta, Actinimycetota, Bacteroidota,* and *Pseudomonadota*– are abundantly present on the human skin (Grice et al. 2009). Various fungal genera are present on the epidermis of a healthy person, and their distribution on the body varies greatly. Nevertheless, in pathological situations, particular species dominate the disturbed area (Cui et al. 2013). The epidermis prevents pathogens from entering the body and acts as a barrier to stop them. Microbes are present on the skin and take shelter there. The variety of microbes depends upon the type of skin a person has. Some microbes reside on glands and cell surfaces (Willey et al. 2011).

Conjunctiva

Microbes exist in small quantities in the conjunctiva. Both gram-positive and gram-negative bacteria are present therein. The lachrymal gland helps hydratethe conjunctiva by secreting the fluid (Cui et al. 2013).

Human Gut

Birth determines the composition of microbes in the human gut (Yang et al. 2016). Delivery influences the quantity of microbes in the stomach. Babies with vaginal birth have a healthy microflora in the gut (Mueller et al. 2015). Babies delivered through C-sections take longer to acquire a beneficial gut microbiome, and dangerous bacteria like *Escherichia coli* and *Staphylococcus aureus* are more prevalent in their gut microbiota (Wall et al. 2009).

Vagina

The numerous microorganisms dwelling inside the vagina constitute the 'vaginal microbiota'. These microbes are essential for keeping the vagina healthy and avoiding infections. *Lactobacilli* species inhibit pathogen development by generating acids and hydrogen peroxide, which is most prevalent among the vaginal bacteria of premenopausal women. Microbial species change with the stages of menstruation (Wang et al. 2014). Heritage has an impact on the vaginal flora as well. The incidence of lactobacilli that create hydrogen peroxide is lesser in American and African women, whereas the pH of their vagina is higher (Antonio et al. 1999).

Placenta

Although it was formerly believed that the placenta is sterile, various microbial species have been discovered there (Wassenaar and Panigrahi 2014). However, the existence of a microbiome in the placenta has been refuted in several studies, raising questions about its existence (De Goffau et al. 2019).

Uterus

Historically, it was believed that the female upper reproductive system was a sterile space. A wide range of microbes live in the uterus of healthy and mature females. The microbiomes of the vagina and the digestive system, and the uterus are highly distinct (Franasiak and Scott Jr 2015).

Oral Cavity

The mouth enhances the growth of common microbes. It provides nutrients and a welcoming environment. Oral microbes cling to the teeth and gums to avoid entering the stomach for protection from acids (Cui et al. 2013). Microbes exist in mouth tissues, creating colonies (Kumar 2013). Saliva is essential in colony formation and biofilm development (Cepoi et al. 2016). It also controls the temperature and provides nutrients (Avila et al. 2009). These microbes can sense their atmosphere and avoid or change their hosts. A powerful inherent host defence mechanism continuously monitors bacterial colonization and prevents it from entering the neighbouring tissues. The host's natural defence mechanism and the dental plaque bacteria coexist in a dynamic equilibrium (Rogers 2008).

Nasal Cavity

The nasal microbiome prevents viral infections (Rhoades et al. 2021).

Lungs

Like the upper and lower respiratory tracts, the oral cavity has barriers to keep out bacteria. Mucus-secreting cells entangle bacteria and drop them out of the respiratory tract. Nasal mucus also has a bactericidal action since lysozyme is present in it. The bacteria in the upper and lower respiratory passages are apparently different (Man et al. 2017). The pulmonary bacterial microbiota comprises nine major bacterial taxa that can cause severe sickness, especially those with weakened immune systems (Cui et al. 2013).

Biliary Tract

In the past, it was believed that the biliary system was non-infectious in a healthy state and that the availability of bacteria indicated sickness. This notion was supported by the inability to allocate bile ducts to bacteria. According to publications that began to appear in 2013, the typical biliary microbiome serves as a distinct functional covering that prevents the colonization of the bile tract by invading microbes (Verdier et al. 2015).

Applications of Probiotics and Diseases Related to the Human Gut Microbiome

Antibiotic-Associated Diarrhea (AAD)

The alteration of the gut flora causes AAD, a frequent side effect of antibiotic therapy. A disease-causing bacteria called *Clostridium difficile* causes AAD, as it can enter the large intestine due to decreased antibiotic resistance (Blaabjerg et al. 2017). Several randomized controlled trials

(RCT) have interestingly revealed that probiotics may be helpful and safe in protecting AAD. Probiotics, for instance, help prevent diarrhea caused by *C. difficile* in adults and children (Conway et al. 2007, Vanderhoof et al. 1999). According to a meta-analysis study, the preventive benefits of probiotics as adjunct treatment may be employed to prevent AAD in outpatients of all ages (Blaabjerg et al. 2017). According to this research, probiotic treatment may reduce AAD incidence by 51% with no discernible rise in adverse effects (Blaabjerg et al. 2017). Additionally, aggregated data from several studies have revealed that *Saccharomyces boulardii* and *Lactobacillus rhamnosus* provide the most significant protection against AAD. Additional research is required to clarify the underlying processes of AAD as produced by bacterial infection,for the precise etiology of the disease has not yet been identified (Kim et al. 2019).

Crohn's Disease (CD)

Abdominal discomfort, diarrhea, fever, exhaustion and weight loss are just a few symptoms that follow CD, an inflammatory bowel disease affecting the whole GIT. Although the exact etiology of CD is unclear, it has been theorized that various elements, including microbiological, genetic and environmental factors, contribute to the disease's development (Baumgart and Carding 2007). The symptoms of CD may be managed in several ways, but there is currently no known treatment. For instance, one can take steroid medicines to minimise intestinal inflammation, and one might use immunosuppressants to lessen immune system activity (Cowan et al. 2010). Probiotics might similarly provide an alternate strategy in addition to traditional treatment. In contrast to late supplementation, Fedorak et al. showed that probiotic administration in CD patients was effective following surgery (Fedorak et al. 2015). Further recent research, however, found that intestinal inflammation remained insignificantly elevated in CD patients after receiving multi-strain probiotic adjuvant therapy (Bjarnason et al. 2019).

Inflammatory Bowel Disease (IBD)

IBD is a persistent digestive tract inflammatory illness. Although the specific origin of IBD is still unclear, it is thought to be caused by an abnormal immune response. IBD is believed to have possible triggers from an unbalanced diet and stress. According to some data, the pathophysiology of IBD may be influenced by the intestinal microbiota. According to several studies, the microbiota makeup differs between healthy people and IBD patients (Shadnoush et al. 2015).

Additionally, it has been hypothesized that controlling the gut microbiota's balance may be vital for avoiding IBD (Yang and Yu 2018).

Probiotics have gained a lot of consideration in the current years as a potential therapy for IBD by altering the microbiota and its positive effects. For example, probiotics have treated ulcerative colitis to cause remission (Furrie et al. 2005). However, according to recent research, probiotic implementation in individuals with IBD is a viable adjuvant therapy for UC but not for CD (Bjarnason et al. 2019). Currently, there is not enough information on the effectiveness of probiotics for us to be able to furnish broad recommendations on their usage on CD patients (Kim et al. 2019).

Colorectal Cancer (CRC)

Colon cancer is a type of cancer that may develop in the colon, the rectum, or any other region of the large intestine. In affluent nations like the United States, Australia and Europe, colon cancer incidence rates have been rising (Center et al. 2009). Bloody feces and weight loss are two signs of this illness. However, hereditary instances are rare and are mainly brought on by age and lifestyle. The prevalence of colon cancer in young people is high in underdeveloped nations, where routine health monitoring is not practicable (Yang and Yu 2018).

Additionally, those with inflammatory bowel diseases like UC or CD are more likely to acquire colorectal cancer (Triantafillidis et al. 2009). It is interesting to note that current research has examined the potential immunomodulatory processes of probiotics in protecting colorectal cancer by altering the gut microbiota. However, as all probiotics do not have the same effects, more research, clinical trials, and animal model research are necessary to determine if probiotic cancer treatment can be successful (Liong 2008). The analysis of clinical studies has revealed that synbiotics cause a considerable alteration in the intestinal microbiota of people with colorectal cancer, with a rise in *Lactobacillus* (Krebs 2016). It has been established that certain colonic resection patients have improved mucosal shape due to enhanced epithelial barrier function. Therefore, consuming synbiotics may potentially change how the intestines' metabolism functions. According to these results, probiotics help treat and prevent intestinal problems, including colorectal cancer, in humans and animals (Marteau et al. 2002).

Probiotics and Microbiome Antiviral Mechanisms

The interactions between hosts, viruses and bacteria still need much more research. Further research is beginning to demonstrate how fascinatingly commensal and probiotic bacteria support the host's defence against viral diseases (Li et al. 2019b). The adaptive immune defences, innate immune defences and the mucus layer are the three main lines of protection that viruses must surpass when they come into contact with mucosal surfaces (such as vaginal, respiratory or GI) (Kumamoto and Iwasaki 2012). According to the study, various commensal and probiotic bacteria may significantly impact each of these lines of defence, which might have important implications for several viral infections. The following were implied in direct and indirect antiviral processes: (1) improved mucosal barrier performance (Lieleg et al. 2012, Schroeder 2019), (2) bacteriocins; the discharge of antiviral, antimicrobial peptides (AMPs); (3) prevention of viral adherence to host cells (Botić et al. 2007, Su et al. 2013); and (4) antiviral modulation (de Vrese et al. 2005, Jounai et al. 2012).

Mucus, a porous biopolymer matrix that covers all wet epithelial surfaces in the human body (including those in the lungs and the vagina), is the main element of the mucosal epithelium barrier (Lieleg et al. 2012). Although it was previously believed that the host is the sole factor controlling the mucus layer, more recent studies have shown that the microbiome also affects its activity and appearance (Schroeder 2019). Pig gastric mucins have been proven to prevent influenza A and human papillomavirus type 16 from invading epithelial cells (Lieleg et al. 2012). The composition of the cervicovaginal mucus also seems to be influenced by the vaginal microbiome. It has been shown that HIV-1 virions disperse more slowly in a microbiota with a *Lactobacillus crispatus* dominance than in a microbiome with a *Lactobacillusiners* dominance or when *Gardnerella vaginalis* is found in high concentrations (Nunn et al. 2015).

Additionally, it has been discovered that a few species, such as *Lacticaseibacillus casei* and *Bifidobacterium adolescentis*, indirectly retain boundary permeability by generating metabolites linked to decreased appearance of the rotavirus toxin NSP4, which disrupts the operation and structure of tight junctions during infection (Gonzalez-Ochoa et al. 2017). Antimicrobial peptides are defence mechanisms in all living organisms, including bacteria and humans (Jenssen and Hamill 2006). All major bacterial lineages generate bacterial AMPs, also termed bacteriocins, which have long been considered essential probiotic characteristics (Dobson et al. 2012). Bacteriocins have been shown to have antimicrobial activity against bacterial infections (Dobson et al. 2012), but it is unknown if they also have antiviral activity. The method of action has also been studied little. The available evidence points to two distinct courses of action. To prevent the Zika virus from infecting the host, it has been shown that the class-1 bacteriocin duramycin produced by *Streptomycetes* inhibits the co-receptor TIM1 (Tabata et al. 2016).

While certain bacteriocins do not obstruct viral entry, they interact with late viral cycle phases to decrease the cytopathic effects and release output. *Enterocin CRL35* from *Enterococcus faecium*, as contrasted to subtilisin, has been shown to reduce glycoprotein synthesis, halting what seems to be an initial step in the HSV infectious cycle (Wachsman et al. 2003). The initial step in virus infection is the attachment of the virus to the host cell (Doms 2016). In a cell culture model, it has been revealed that some strains of *Bifidobacterium* and *Lactobacillus* stop the binding and penetration of the vesicular stomatitis virus into cells, perhaps through steric hindrance (Botić et al. 2007).

The term 'innate immune system' describes initial, non-specific defensive responses towards exposure to viral antigens. Physical obstructions, such as antimicrobial peptides, innate leucocytes, cellular receptors, epithelial cell surfaces and phagocytes, are all parts of this system. Pattern-recognition receptors (PRRs) on epithelial cells, dendritic cells (DCs) and macrophages continuously survey the mucosal environment and identify the occurrence of attacking viruses. It is known that NK cells are essential for viral infection in humans. Probiotics dramatically boost NK cell activity in healthy older people, according to a recent analysis of six RCTs with different strains and doses (Gui et al. 2020).

Viral inactivation by LAB may potentially take place through an adsorption or trapping method (Al Kassaa et al. 2014). The antiviral benefits of probiotics are shown using this technique, depending on the interaction of bacteria and viruses (Karthik et al. 2014). Researchers claim that *Lactobacillus* isolated from the vagina has a protectiveeffect against HSV via the bacteria's capability to serve as an adhesive, which reduces the absorption of the virus. Direct interactions between viral particles and bacterial cells may result in entrapment or competition for cell membrane receptors (Vickers 2017). According to a prior investigation, the probiotics interact directly with the virus's envelope to capture the vesicular stomatitis virus (Indiana vesiculovirus) (Botić et al. 2007). Additionally, Wang et al. (2013) found that *Enterococcus faecium* directly interacts with influenza viruses to suppress them. However, these processes may stop SARS-CoV-2 infection or reduce viral load, impacting the severity of the disease (Salman et al. 2020).

Probiotics, the Microbiome, and SARS-CoV-2

Angiotensin-converting enzyme 2 (ACE2) receptors are essential for SARS-CoV-2 to bind to and invade human cells (Wu et al. 2020). According to a recent meta-analysis, more than 50% of the individuals with COVID-19 had SARS-CoV-2 mRNA in their feces, with the detection rate being notably greater in severe forms and in those exhibiting GI symptoms. It has been shown that enterocyte organoids experience productive infection *in vitro* (Lamers et al. 2020). Vomiting seems more common in children, but diarrhea is the common GI indication in adults and children (Cheung et al. 2020).

It should be noted that viral mRNA detection in the stool may persist in specific individuals even after respiratory sample results are negative (Wu et al. 2020). Representatives of the genera *Bifidobacterium* and *Lactobacillus*are less in an initial restricted case series of COVID-19 disease individuals in China (Chen et al. 2020). These results are supported methodologically by including hospitalized patients with different diseases as controls in these trials, and one of the studies additionally took into account antibiotic co-therapy.

However, it is still unknown if microbiome changes contribute to the severity of COVID-19. Whatever advantages probiotics may have, it is doubtful they will directly affect SARS-CoV-2 infection. In this regard, probiotics may help repair the microbiota imbalances seen in certain COVID-19 instances by preventing the development of the specific opportunistic bacteria in question and/or promoting the recovery of good bacteria. Additionally, probiotics have been proposed to alleviaterespiratory tract infections. This mechanism involves direct interactions between immune cells and probiotics and the reduction of intestinal permeability (Baud et al. 2020), as shown in Figure 2. The metabolism of tryptophan significantly impacts mucosal immunity via the kynurenine, serotonin and indole pathways, all of which are controlled by the microbiota (Agus et al. 2018). This process also has several additional systemic effects. As a result, particular microbiome therapies in COVID-19 may have tryptophan-dependent pathwaysas their target.

Viral Infections

The WHO declared a global public health emergency and pandemic on March 11, 2020, in response to the growing danger of SARS-CoV-2. Coronavirus illness 19,brought on by SARS-CoV-2 infection, is accompanied by fever, respiratory and gastrointestinal signs, and other less typical ones (Cheung et al. 2020). SARS-CoV-2 still lacks an authorized vaccine or antiviral medication. However, new solutions and approaches to these issues are required in light of this epidemic. It has been generally documented that boosting the immune system is an efficient and beneficial strategy for leading a healthy lifestyle. The need for a robust immune system to combat SARS-CoV-2 infection was thus brought to light (Arshad et al. 2020). An open literature search found that a nutritious diet significantly enhances immune system performance. Therefore, a varied diet with a wide range of nutrients may help avoid or even lessenour susceptibility to COVID-19. Healthy dietary options, vitamins, bioactive substances and probiotics may be adjuvants against COVID-19 (de Araújo Morais et al. 2021). The potential advantages of probiotics in treating other coronavirus strains have been amply supported by research (Chai et al. 2013). Due to their antiviral action, capacity to reduce inflammation, ability to repair the gut flora, quick accessibility, general safety, low cost, and ease of administration, certain probiotics may also be useful in the treatment of COVID-19 patients (Angurana and Bansal 2021). However, there are no recommendations for probiotic strains, dosage or duration. Nonetheless, *Bifidobacterium* and *Lactobacillus*may be utilized safely in many therapeutic settings.

The antiviral action of plantaricins, a bacteriocin produced as a metabolic byproduct of the bacterium *L. plantarum*, has been studied inthe latest research on computational and molecular dynamics by Aanouz et al. (2021). One specific therapy approach for preventing cytokine storm that may be proposed in addition to boosting the immune system is using probiotics for individuals with gastrointestinal symptoms related to COVID-19 and those with less severe systemic symptoms (Infusino et al. 2020).

How different patients react to infections may vary depending on the makeup of their microbiota, which may be corrected with probiotics to reduce the intensive care requirement (Inchingolo et al. 2018). The possible contribution of the gut microbiota to improved immunological and respiratory function in these individuals has been noted by Dhar and Mohanty (2020). Additionally, several writers have reported that supplementing with probiotics and/or synbiotics might significantly improve respiratory performance, building on earlier research on the strong correlations between gut microbiota and lung function. This research confirms the importance of probiotics as supplements for patients and the part our microbiota plays in maintaining as well as regaining a good quality of life. Probiotics benefit humans in many ways, but they may also play a role in specific clinical situations that might be lethal, such as bloodstream infections (Ventoulis et al. 2020, Vickers 2017). However, using probiotics for therapeutic purposes has not yet been associated with any deaths or severe health concerns.

Probiotics and Intestinal Neuroimmunology

The central nervous system (CNS) and the gut interact with each other in both gut-to-brain and brain-to-gut directions, and this interaction is modulated by the enteric nervous system, a vast and complex neural network found in the human gastrointestinal tract (Sharma et al. 2009). The enteric nervous system is primarily in charge of controlling the physiological functions of the gut. A network of intricate reflex loops in the nervous system called the gut-brain axis, serves as a channel forcommunication (Mayer 2011). The brain-gut axis controls the flow of information between the brain, the gastrointestinal tract, and the immunological and endocrine schemes associated with sustaining gut purpose. According to Neufeld and Foster (2009), disruptions or abnormalities in the gut-brain axis are responsible for functional gastrointestinal disorders like IBS and psychological symptoms like anxiety (Bienenstock and Collins 2010).

Researchers now have unprecedented chances to examine the human microbiome's structure and role and its relationship with neurological illnesses because of research into the human microbiome and the advancement of next-generation sequencing tools. In intestinal blockade dysfunction, variations in the inflammation and the gut microbiota have been linked to hepatic encephalopathy, a disease often seen in cirrhotic individuals with liver failure and defined by impairments in cognitive processes (Bajaj et al. 2012). Current scientific research has provided insights into the complex connections between the gut-brain axis and the intestinal microbiota. By producing neuroactive and neuroendocrine chemicals, including adrenaline, GABA, noradrenaline, serotonin and histamine, gut microorganisms may be able to connect with the gut-brain axis (Bienenstock et al. 2010, Forsythe et al. 2010). While direct data does not show that intestinal bacteria produce serotonin, a metabolomic investigation utilizing germ-free mice has shown that conventional animals have blood serotonin levels that are 2.8 times greater than those of germ-free mice (Wikoff et al. 2009). GABA, a neurotransmitter that acts as a repressive neurotransmitter in the CNS and may be involved in regulating pain, may be produced by other intestinal microorganisms, such as *Lactobacilli* (Li and Cao 2010, Su et al. 2011). In a colorectal distension model using Sprague-Dawley rats, the suppressive impact of gut bacteria on visceral discomfort originating from the gastrointestinal system wasshown by researchers. When colonic distension is used, 9 days of *L. rhamnosus ATCC23272* therapy decreases the rats' perception of discomfort (Kamiya et al. 2006).

Intestinal bacteria and the enteric nervous system may work in tandem to influence behaviour, pain perception, and immune responses in the gut and extraintestinal sites. Jarchum and Pamer claim in their exhaustive analysis published in 2011 that gut bacteria have a role in immune system development and homeostasis (Jarchum and Pamer 2011). The opportunity for gut microbes to participate in the expansion of neuroimmunological diseases is intriguing (Ochoa-Repáraz et al. 2011). According to studies using a mouse model of experimental autoimmune encephalomyelitis (EAE), the gut microbiota may act as a catalyst for autoimmunity mediated by myelin-specific CD4+ T cells. Antibiotics and probiotics like *L. paracasei* DSM 13434 and *L. plantarum* DSM 15312 have been used to treat EAE clinical indications and reduce inflammation (Ochoa-Repáraz et al. 2009, Yokote et al. 2008). These findings suggest that probiotics may alter the gut microbiota and benefit those suffering from autoimmune illnesses.

Probiotics and Viral Infections

Probiotic bacteria often come from the genera *Bifidobacterium* and *Lactobacillus*. These organisms are widely used to treat inflammations, infectious diarrhea, atopic dermatitis, IBD and migraines. However, their actual functioning is more complicated since it has been demonstrated that they may elicit both pro- and anti-inflammatory responses (Chiba et al. 2010, Plaza-Diaz et al. 2013). Probiotics, prebiotics and postbiotics, as well as nonviable microbial cells, microbial fractions and cell lysates (Aguilar-Toalá et al. 2018), are examples of therapies that work on the microbiome and have a long history of usage for a variety of diseases.

Researchers do not yet completely understand the processes by which certain probiotic strains may cause these antiviral effects. But there are theories about how these systems work. The antiviral effects of probiotics may manifest in several ways, such as the synthesis of chemicals that hinder the growth of viruses directly, immune system stimulation, or direct contact with viruses (Al Kassaa et al. 2014). Probiotics may advance the activity of the mucosal barrier, comparable to the endogenous microbiota, thus reducing the capacity of viral particles to cross this physical barrier and sustaining this blockage throughout the viral disease (Bron et al. 2017). Direct virucidal effects have been shown in a subset of probiotic lactobacilli, mediated by primary metabolites such as hydrogen peroxide or organic acids that lower the pH (Zabihollahi et al. 2012). In addition to their well-known antibacterial effects, certain bacteriocins have also been shown to have antiviral action (Quintana et al. 2014, Torres et al. 2013). It has been shown that, among other species, adenoviruses are effectively blocked by exopolysaccharides produced by lactic acid bacteria during development (Biliavska et al. 2019).

Furthermore, it has been shown that *Levilactobacillusbrevis* cell wall fragments decrease the replication capacity of herpes simplex, and other metabolites and bacterial cell fragments may block the binding of viruses (Aktories et al. 2018, Martín et al. 2010). Nevertheless, the probiotic effect on the immune system has received the greatest attention in research on its antiviral capabilities. Increasing cytotoxic and NK cell activity has been related to several probiotics (Miller et al. 2019). The antibody and antiviral cytokine responses, like IFN-g, IL-18, IL-12 and IL-2, have improved

(Lehtoranta et al. 2020). Many of these and other operations might run simultaneously or sequentially.

Probiotics were initially classified as having antiviral properties in addition to their significant biological functions. Research by Ang et al. (2016) has shown that *Lactobacillus reuteri* may protect human skeletal muscles and colonic cell lines against Coxsackievirus A and Enterovirus 71 infections. Furthermore, *Lactobacillus casei* and *Bifidobacterium adolescentis* have been shown to have antiviral action against rotavirus infection (Olaya Galán et al. 2016). Moreover, *Bacillus subtilis* has been demonstrated to have anti-influenza action. On the other hand, *Bacillus subtilis* and *Lactobacillus gasseri* have been shown to have potential against sentential respiratory viruses (Starosila et al. 2017, Eguchi et al. 2019). However, further research is still needed to determine how beneficial probiotics are in treating certain viral infections (Salman et al. 2020).

Conclusion and Future Direction

Current studies on the structure and operation of the microbiome suggest that diet may directly affect the intestinal microbiota and the health of humans and animals. Disruptions in the relationships between microbes and humans may lead to various disease states, such as chronic inflammation, autoimmune diseases and neurological disorders. In this light, probiotics have been recommended as a therapeutic and preventive strategy to assist the gut microbiome in regaining its normalcomposition and functionality. Also, information from research on the human microbiome may help identify new native microbial species and develop techniques to influence changes in the microbial populations in the gut. *In vitro* or *in vivo* investigations using suitable and well-designed experimental models may provide information on the biology and possible manipulation of the human host's microbiome. Probiotics, intestinal bacteria and the human gastrointestinal system can all interact with one another. This can be investigated worldwide using metagenomic, metatranscriptomic and metabonomics techniques. Future approaches to enhance health, prevent illness and cure various ailments may use new varieties of probiotics or pharmaceutical substances produced from the microbiome. Probiotics have shown promise in treating disorders linked to the human gut microbiome in earlier research assessing human intestinal microbiota post-consumption.

Additionally, research into identifying and assessing the contributions of probiotics to human gastrointestinal disorders has intensified recently. Probiotics may be utilized as treatments or preventive measures for human gut microbiome-related disorders. However, more hypothesis-driven investigations are still required.

Acknowledgments

The authors acknowledge the support of the Cholistan University of Veterinary and Animal Sciences-Bahawalpur, Pakistan, for this chapter.

Authors' Contributions

Kamal Niaz and Wen-Jun Li conceived the original idea and designed the outlines of the study. Firasat Hussain, Shafeeq Ur Rehman and Muhammad Naveed Nawaz equally contributed and wrote the first draft of the manuscript. Ahmed Abdelmoneim and Kamal Niaz revised the whole manuscript and formatted it accordingly. Ahmed Abdelmoneim, Kamal Niaz, Wen-Jun Li and Kashif Rahim critically edited the final draft. All the authors have read and approved the final manuscript.

References

Aanouz, I., Belhassan, A., El-Khatabi, K., Lakhlifi, T., El-Ldrissi, M. and Bouachrine, M. (2021). Moroccan medicinal plants as inhibitors against SARS-CoV-2 main protease: Computational investigations. Journal of Biomolecular Structure and Dynamics, 39: 2971–2979.

Aguilar-Toalá, J., Garcia-Varela, R., Garcia, H., Mata-Haro, V., González-Córdova, A., Vallejo-Cordoba, B. and Hernández-Mendoza, A. (2018). Postbiotics: An evolving term within the functional foods field. Trends in Food Science & Technology, 75: 105–114.

Agus, A., Planchais, J. and Sokol, H. (2018). Gut microbiota regulation of tryptophan metabolism in health and disease. Cell Host & Microbe, 23: 716–724.

Aktories, K., Papatheodorou, P. and Schwan, C. (2018). Binary Clostridium difficile toxin (CDT)-A virulence factor disturbing the cytoskeleton. Anaerobe, 53: 21–29.

Al Kassaa, I., Hober, D., Hamze, M., Chihib, N. E. and Drider, D. (2014). Antiviral potential of lactic acid bacteria and their bacteriocins. Probiotics and Antimicrobial Proteins, 6: 177–185.

Alander, M., Korpela, R., Saxelin, M., Vilpponen-Salmela, T., Mattila-Sandholm, T. and Von Wright, A. (1997). Recovery of Lactobacillus rhamnosus GG from human colonic biopsies. Letters in Applied Microbiology, 24: 361–364.

Allen, S. J., Martinez, E. G., Gregorio, G. V. and Dans, L. F. (2010). Probiotics for treating acute infectious diarrhoea. Cochrane Database of Systematic Reviews.

Amar, J., Burcelin, R., Ruidavets, J. B., Cani, P. D., Fauvel, J., Alessi, M. C., Chamontin, B. and Ferriéres, J. (2008). Energy intake is associated with endotoxemia in apparently healthy men. The American Journal of Clinical Nutrition, 87: 1219–1223.

Ang, L. Y. E., Too, H. K. I., Tan, E. L., Chow, T. -K. V., Shek, P. -C. L., Tham, E. and Alonso, S. (2016). Antiviral activity of Lactobacillus reuteri Protectis against Coxsackievirus A and Enterovirus 71 infection in human skeletal muscle and colon cell lines. Virology Journal, 13: 1–12.

Angurana, S. K. and Bansal, A. (2021). Probiotics and Coronavirus disease 2019: Think about the link. British Journal of Nutrition, 126: 1564–1570.

Antonio, M. A., Hawes, S. E. and Hillier, S. L. (1999). The identification of vaginal Lactobacillus species and the demographic and microbiologic characteristics of women colonized by these species. Journal of Infectious Diseases, 180: 1950–1956.

Apajalahti, J. (2005). Comparative gut microflora, metabolic challenges, and potential opportunities. Journal of Applied Poultry Research, 14: 444–453.

Arshad, M. S., Khan, U., Sadiq, A., Khalid, W., Hussain, M., Yasmeen, A., Asghar, Z. and Rehana, H. (2020). Coronavirus disease (COVID-19) and immunity booster green foods: A mini review. Food Science & Nutrition, 8: 3971–3976.

Avena-Woods, C. (2017). Overview of atopic dermatitis. The American Journal of Managed Care, 23: S115–S123.

Avila, M., Ojcius, D. M. and Yilmaz, Ö. (2009). The oral microbiota: Living with a permanent guest. DNA and Cell Biology, 28: 405–411.

Bae, J. -Y., Kim, J. I., Park, S., Yoo, K., Kim, I. -H., Joo, W., Ryu, B. H., Park, M. S., Lee, I. and Park, M. -S. (2018). Effects of Lactobacillus plantarum and Leuconostoc mesenteroides probiotics on human seasonal and avian influenza viruses.

Bagarolli, R. A., Tobar, N., Oliveira, A. G., Araújo, T. G., Carvalho, B. M., Rocha, G. Z., Vecina, J. F., Calisto, K., Guadagnini, D. and Prada, P. O. (2017). Probiotics modulate gut microbiota and improve insulin sensitivity in DIO mice. The Journal of Nutritional Biochemistry, 50: 16–25.

Bajaj, J. S., Ridlon, J. M., Hylemon, P. B., Thacker, L. R., Heuman, D. M., Smith, S., Sikaroodi, M. and Gillevet, P. M. (2012). Linkage of gut microbiome with cognition in hepatic encephalopathy. American Journal of Physiology-Gastrointestinal and Liver Physiology, 302: G168–G175.

Baud, D., Dimopoulou Agri, V., Gibson, G. R., Reid, G. and Giannoni, E. (2020). Using probiotics to flatten the curve of coronavirus disease COVID-2019 pandemic. Frontiers in Public Health, 8: 186.

Baumgart, D. C. and Carding, S. R. (2007). Inflammatory bowel disease: Cause and immunobiology. The Lancet, 369: 1627–1640.

Bienenstock, J. and Collins, S. (2010). 99th Dahlem conference on infection, inflammation and chronic inflammatory disorders: Psycho-neuroimmunology and the intestinal microbiota: Clinical observations and basic mechanisms. Clinical & Experimental Immunology, 160: 85–91.

Bienenstock, J., Forsythe, P., Karimi, K. and Kunze, W. (2010). Neuroimmune aspects of food intake. International Dairy Journal, 20: 253–258.

Biliavska, L., Pankivska, Y., Povnitsa, O. and Zagorodnya, S. (2019). Antiviral activity of exopolysaccharides produced by lactic acid bacteria of the genera Pediococcus, Leuconostoc and Lactobacillus against human adenovirus type 5. Medicina, 55: 519.

Bjarnason, I., Sission, G. and Hayee, B. H. (2019). A randomised, double-blind, placebo-controlled trial of a multi-strain probiotic in patients with asymptomatic ulcerative colitis and Crohn's disease. Inflammopharmacology, 27: 465–473.

Blaabjerg, S., Artzi, D. M. and Aabenhus, R. (2017). Probiotics for the prevention of antibiotic-associated diarrhea in outpatients—A systematic review and meta-analysis. Antibiotics, 6: 21.

Botić, T., Danø, T., Weingartl, H. and Cencič, A. (2007). A novel eukaryotic cell culture model to study antiviral activity of potential probiotic bacteria. International Journal of Food Microbiology, 115: 227–234.

Bron, P. A., Kleerebezem, M., Brummer, R. -J., Cani, P. D., Mercenier, A., MacDonald, T. T., Garcia-Ródenas, C. L. and Wells, J. M. (2017). Can probiotics modulate human disease by impacting intestinal barrier function? British Journal of Nutrition, 117: 93–107.

Brookes, M. J., Woolrich, M., Luckhoo, H., Price, D., Hale, J. R., Stephenson, M. C., Barnes, G. R., Smith, S. M. and Morris, P. G. (2011). Investigating the electrophysiological basis of resting state networks using magnetoencephalography. Proceedings of the National Academy of Sciences, 108: 16783–16788.

Caballero, S., Carter, R., Ke, X., Sušac, B., Leiner, I. M., Kim, G. J., Miller, L., Ling, L., Manova, K. and Pamer, E. G. (2015). Distinct but spatially overlapping intestinal niches for vancomycin-resistant Enterococcus faecium and carbapenem-resistant Klebsiella pneumoniae. PLoS Pathogens, 11: e1005132.

Caffarelli, C., Cardinale, F., Povesi-Dascola, C., Dodi, I., Mastrorilli, V. and Ricci, G. (2015). Use of probiotics in pediatric infectious diseases. Expert Review of Anti-infective Therapy, 13: 1517–1535.

Cani, P. and de Vos, W. (2017). Next-generation beneficial microbes: The case of Akkermansia muciniphila. Front. Microbiol., 8: 1765.

Cani, P. D., Bibiloni, R., Knauf, C., Waget, A., Neyrinck, A. M., Delzenne, N. M. and Burcelin, R. (2008). Changes in gut microbiota control metabolic endotoxemia-induced inflammation in high-fat diet–induced obesity and diabetes in mice. Diabetes, 57: 1470–1481.

Carding, S., Verbeke, K., Vipond, D. T., Corfe, B. M. and Owen, L. J. (2015). Dysbiosis of the gut microbiota in disease. Microbial Ecology in Health and Disease, 26: 26191.

Center, M. M., Jemal, A., Smith, R. A. and Ward, E. (2009). Worldwide variations in colorectal cancer. CA: A Cancer Journal for Clinicians, 59: 366–378.

Cepoi, L., Donţu, N., Şalaru, V. and Şalaru, V. (2016). Removal of organic pollutants from wastewater by cyanobacteria. Cyanobacteria for Bioremediation of Wastewaters, Springer, pp. 27–43.

Cha, B. K., Jung, S. M., Choi, C. H., Song, I. -D., Lee, H. W., Kim, H. J., Hyuk, J., Chang, S. K., Kim, K. and Chung, W. -S. (2012). The effect of a multispecies probiotic mixture on the symptoms and fecal microbiota in diarrhea-dominant irritable bowel syndrome: A randomized, double-blind, placebo-controlled trial. Journal of Clinical Gastroenterology, 46: 220–227.

Chai, W., Burwinkel, M., Wang, Z., Palissa, C., Esch, B., Twardziok, S., Rieger, J., Wrede, P. and Schmidt, M. F. (2013). Antiviral effects of a probiotic Enterococcus faecium strain against transmissible gastroenteritis coronavirus. Archives of Virology, 158: 799–807.

Chen, J., Liu, D., Liu, L., Liu, P., Xu, Q., Xia, L., Ling, Y., Huang, D., Song, S., Zhang, D. and Qian, Z. (2020). A pilot study of hydroxychloroquine in treatment of patients with moderate COVID-19. Zhejiang da xue xue bao. Yi xue ban= Journal of Zhejiang University. Medical Sciences, 49(2): 215–219.

Cheung, K. S., Hung, I. F., Chan, P. P., Lung, K., Tso, E., Liu, R., Ng, Y., Chu, M. Y., Chung, T. W. and Tam, A. R. (2020). Gastrointestinal manifestations of SARS-CoV-2 infection and virus load in fecal samples from a Hong Kong cohort: systematic review and meta-analysis. Gastroenterology, 159: 81–95.

Chiba, Y., Shida, K., Nagata, S., Wada, M., Bian, L., Wang, C., Shimizu, T., Yamashiro, Y., Kiyoshima-Shibata, J. and Nanno, M. (2010). Well-controlled proinflammatory cytokine responses of Peyer's patch cells to probiotic Lactobacillus casei. Immunology, 130: 352–362.

Choi, S., Hwang, Y. -J., Shin, M. -J. and Yi, H. (2017). Difference in the gut microbiome between ovariectomy-induced obesity and diet-induced obesity. Journal of Microbiology and Biotechnology, 27: 2228–2236.

Clemente, J. C., Ursell, L. K., Parfrey, L. W. and Knight, R. (2012). The impact of the gut microbiota on human health: An integrative view. Cell, 148: 1258–1270.

Conway, S., Hart, A., Clark, A. and Harvey, I. (2007). Does eating yogurt prevent antibiotic-associated diarrhoea?: A placebo-controlled randomised controlled trial in general practice. British Journal of General Practice, 57: 953–959.

Cowan, D. C., Cowan, J. O., Palmay, R., Williamson, A. and Taylor, D. R. (2010). Effects of steroid therapy on inflammatory cell subtypes in asthma. Thorax, 65: 384–390.

Cremonini, F., Di Caro, S., Nista, E. C., Bartolozzi, F., Capelli, G., Gasbarrini, G. and Gasbarrini, A. (2002). Meta-analysis: The effect of probiotic administration on antibiotic-associated diarrhoea. Alimentary Pharmacology & Therapeutics, 16: 1461–1467.

Cui, L., Morris, A. and Ghedin, E. (2013). The human mycobiome in health and disease. Genome Medicine, 5: 1–12.

Dairi, T. (2012). Menaquinone biosyntheses in microorganisms. Methods in Enzymology, vol. 515 Elsevier, pp. 107–122.

de Araújo Morais, A. H., de Souza Aquino, J., da Silva-Maia, J. K., de Lima Vale, S. H., Maciel, B. L. L. and Passos, T. S. (2021). Nutritional status, diet and viral respiratory infections: Perspectives for severe acute respiratory syndrome coronavirus 2. British Journal of Nutrition, 125: 851–862.

De Biase, D. and Pennacchietti, E. (2012). Glutamate decarboxylase-dependent acid resistance in orally acquired bacteria: function, distribution and biomedical implications of the gadBC operon. Molecular Microbiology, 86: 770–786.

De Filippo, C., Cavalieri, D., Di Paola, M., Ramazzotti, M., Poullet, J. B., Massart, S., Collini, S., Pieraccini, G. and Lionetti, P. (2010). Impact of diet in shaping gut microbiota revealed by a comparative study in children from Europe and rural Africa. Proceedings of the National Academy of Sciences, 107: 14691–14696.

De Goffau, M. C., Lager, S., Sovio, U., Gaccioli, F., Cook, E., Peacock, S. J., Parkhill, J., Charnock-Jones, D. S. and Smith, G. C. (2019). Human placenta has no microbiome but can contain potential pathogens. Nature, 572: 329–334.

de Vrese, M., Rautenberg, P., Laue, C., Koopmans, M., Herremans, T. and Schrezenmeir, J. (2005). Probiotic bacteria stimulate virus–specific neutralizing antibodies following a booster polio vaccination. European Journal of Nutrition, 44: 406–413.

Depoorter, L. and Vandenplas, Y. (2022). Probiotics in pediatrics. Probiotics, 425–450.

Derrien, M. and van Hylckama Vlieg, J. E. (2015). Fate, activity, and impact of ingested bacteria within the human gut microbiota. Trends in Microbiology, 23: 354–366.

Dewulf, E. M., Cani, P. D., Claus, S. P., Fuentes, S., Puylaert, P. G., Neyrinck, A. M., Bindels, L. B., de Vos, W. M., Gibson, G. R. and Thissen, J. -P. (2013). Insight into the prebiotic concept: lessons from an exploratory, double blind intervention study with inulin-type fructans in obese women. Gut, 62: 1112–1121.

Dhar, D. and Mohanty, A. (2020). Gut microbiota and Covid-19-possible link and implications. Virus Research, 285: 198018.

Dobson, A., Cotter, P. and Ross, R. (2012). Production of bacteria: Probiotic trait? Appl. Environ. Microbiol., 78: 1–6.

Doms, R. (2016). Viral Pathogenesis: From Basics to Systems Biology: Academic Press, London.

Duncan, S. H., Louis, P. and Flint, H. J. (2007). Cultivable bacterial diversity from the human colon. Letters in Applied Microbiology, 44: 343–350.

Duparc, T., Plovier, H., Marrachelli, V. G., Van Hul, M., Essaghir, A., Ståhlman, M., Matamoros, S., Geurts, L., Pardo-Tendero, M. M. and Druart, C. (2017). Hepatocyte MyD88 affects bile acids, gut microbiota and metabolome contributing to regulate glucose and lipid metabolism. Gut, 66: 620-632.

Everard, A., Geurts, L., Caesar, R., Van Hul, M., Matamoros, S., Duparc, T., Denis, R. G., Cochez, P., Pierard, F. and Castel, J. (2014). Intestinal epithelial MyD88 is a sensor switching host metabolism towards obesity according to nutritional status. Nature Communications, 5: 1–12.

Falony, G., Vieira-Silva, S. and Raes, J. (2015). Microbiology meets big data: The case of gut microbiota–derived trimethylamine. Annual Review of Microbiology, 69: 305–321.

Fang, K., Jin, X. and Hong, S. (2018). Probiotic Escherichia coli inhibits biofilm formation of pathogenic *E. coli* via extracellular activity of DegP. Sci Rep., 8(1): 4939. Epub 2018/03/23. https://doi. org/10.1038/s41598-018-23180-1 PMID: 29563542. Report no.

Fedorak, R. N., Feagan, B. G., Hotte, N., Leddin, D., Dieleman, L. A., Petrunia, D. M., Enns, R., Bitton, A., Chiba, N. and Paré, P. (2015). The probiotic VSL# 3 has anti-inflammatory effects and could reduce endoscopic recurrence after surgery for Crohn's disease. Clinical Gastroenterology and Hepatology, 13: 928–935. e922.

Feizizadeh, S., Salehi-Abargouei, A. and Akbari, V. (2014). Efficacy and safety of Saccharomyces boulardii for acute diarrhea. Pediatrics, 134: e176–e191.

Flint, H. J., Scott, K. P., Duncan, S. H., Louis, P. and Forano, E. (2012). Microbial degradation of complex carbohydrates in the gut. Gut Microbes, 3: 289–306.

Florin, T. H., Zhu, G., Kirk, K. M. and Martin, N. G. (2000). Shared and unique environmental factors determine the ecology of methanogens in humans and rats. The American Journal of Gastroenterology, 95: 2872–2879.

Forsythe, P., Sudo, N., Dinan, T., Taylor, V. H. and Bienenstock, J. (2010). Mood and gut feelings. Brain, Behavior, and Immunity, 24: 9–16.

Franasiak, J. M. and Scott, Jr. R. T. (2015). Reproductive tract microbiome in assisted reproductive technologies. Fertility and Sterility, 104: 1364–1371.

Freedman, S. B., Williamson-Urquhart, S., Farion, K. J., Gouin, S., Willan, A. R., Poonai, N., Hurley, K., Sherman, P. M., Finkelstein, Y. and Lee, B. E. (2018). Multicenter trial of a combination probiotic for children with gastroenteritis. New England Journal of Medicine, 379: 2015–2026.

Furrie, E., Macfarlane, S., Kennedy, A., Cummings, J., Walsh, S., O'neil, D. and Macfarlane, G. (2005). Synbiotic therapy (Bifidobacterium longum/Synergy 1) initiates resolution of inflammation in patients with active ulcerative colitis: A randomised controlled pilot trial. Gut, 54: 242–249.

Gibson, G. R., Beatty, E. R., Wang, X. and Cummings, J. H. (1995). Selective stimulation of bifidobacteria in the human colon by oligofructose and inulin. Gastroenterology, 108: 975–982.

Gibson, G. R., Hutkins, R., Sanders, M. E., Prescott, S. L., Reimer, R. A., Salminen, S. J., Scott, K., Stanton, C., Swanson, K. S. and Cani, P. D. (2017). Expert consensus document: The International Scientific Association for Probiotics and Prebiotics (ISAPP) consensus statement on the definition and scope of prebiotics. Nature Reviews Gastroenterology & Hepatology, 14: 491–502.

Gonzalez-Ochoa, G., Flores-Mendoza, L. K., Icedo-Garcia, R., Gomez-Flores, R. and Tamez-Guerra, P. (2017). Modulation of rotavirus severe gastroenteritis by the combination of probiotics and prebiotics. Archives of Microbiology, 199: 953–961.

Grice, E. A., Kong, H. H., Conlan, S., Deming, C. B., Davis, J., Young, A. C., Program, N. C. S., Bouffard, G. G., Blakesley, R. W. and Murray, P. R. (2009). Topographical and temporal diversity of the human skin microbiome. Science, 324: 1190–1192.

Guarino, A., Ashkenazi, S., Gendrel, D., Vecchio, A. L., Shamir, R. and Szajewska, H. (2014). European Society for Pediatric Gastroenterology, Hepatology, and Nutrition/ European Society for Pediatric Infectious Diseases evidence-based guidelines for the management of acute gastroenteritis in children in Europe: Update 2014. Journal of Pediatric Gastroenterology and Nutrition, 59: 132–152.

Guarner, F., Khan, A. G., Garisch, J., Eliakim, R., Gangl, A., Thomson, A., Krabshuis, J., Lemair, T., Kaufmann, P. and De Paula, J. A. (2012). World gastroenterology organisation global guidelines: Probiotics and prebiotics October 2011. Journal of Clinical Gastroenterology, 46: 468–481.

Guarner, F., Sanders, M., Eliakim, R., Fedorak, R., Gangl, A. and Garisch, J. (2017). World Gastroenterology Organisation. Probiotics and Prebiotics. February 2017 [homepage en Internet]; 2017 [citado 10 de agosto de 2017].

Guevara-Romero, E., Flórez-García, V., Egede, L. E. and Yan, A. (2022). Factors associated with the double burden of malnutrition at the household level: A scoping review. Critical Reviews in Food Science and Nutrition, 62: 6961–6972.

Gui, Q., Wang, A., Zhao, X., Huang, S., Tan, Z., Xiao, C. and Yang, Y. (2020). Effects of probiotic supplementation on natural killer cell function in healthy elderly individuals:

A meta-analysis of randomized controlled trials. European Journal of Clinical Nutrition, 74: 1630–1637.

Gummesson, A., Carlsson, L. M., Storlien, L. H., Bäckhed, F., Lundin, P., Löfgren, L., Stenlöf, K., Lam, Y. Y., Fagerberg, B. and Carlsson, B. (2011). Intestinal permeability is associated with visceral adiposity in healthy women. Obesity, 19: 2280–2282.

Hannigan, G. D., Meisel, J. S., Tyldsley, A. S., Zheng, Q., Hodkinson, B. P., SanMiguel, A. J., Minot, S., Bushman, F. D. and Grice, E. A. (2015). The human skin double-stranded DNA virome: Topographical and temporal diversity, genetic enrichment, and dynamic associations with the host microbiome. MBio, 6: e01578–01515.

Hill, C., Guarner, F., Reid, G., Gibson, G. R., Merenstein, D. J., Pot, B., Morelli, L., Canani, R. B., Flint, H. J. and Salminen, S. (2014). The international scientific association for probiotics and prebiotics consensus statement on the scope and appropriate use of the term probiotic. Nature Reviews Gastroenterology & Hepatology, 11: 506–514.

Huang, R., Ning, H., Shen, M., Li, J., Zhang, J. and Chen, X. (2017). Probiotics for the treatment of atopic dermatitis in children: A systematic review and meta-analysis of randomized controlled trials. Frontiers in Cellular and Infection Microbiology, 7: 392.

Inchingolo, F., Dipalma, G., Cirulli, N., Cantore, S., Saini, R., Altini, V., Santacroce, L., Ballini, A. and Saini, R. (2018). Microbiological results of improvement in periodontal condition by administration of oral probiotics. Journal of Biological Regulators and Homeostatic Agents, 32: 1323–1328.

Infusino, F., Marazzato, M., Mancone, M., Fedele, F., Mastroianni, C., Severino, P. and d'Ettorre, G. (2020). Diet supplementation, probiotics, and nutraceuticals in SARS-CoV-2 infection: A scoping review: Nutrients.

Iqbal, Z., Ahmed, S., Tabassum, N., Bhattacharya, R. and Bose, D. (2021). Role of probiotics in prevention and treatment of enteric infections: a comprehensive review. 3 Biotech., 11: 1–26.

Jafarnejad, S., Shab-Bidar, S., Speakman, J. R., Parastui, K., Daneshi-Maskooni, M. and Djafarian, K. (2016). Probiotics reduce the risk of antibiotic-associated diarrhea in adults (18–64 Years) but not the elderly (> 65 Years) a meta-analysis. Nutrition in Clinical Practice, 31: 502–513.

Jarchum, I. and Pamer, E. G. (2011). Regulation of innate and adaptive immunity by the commensal microbiota. Current Opinion in Immunology, 23: 353–360.

Jenssen, H. and Hamill, P. (2006). Hancock reW//clin. Microbiol. Rev., 19: 491–511.

Johansson, M. -L., Nobaek, S., Berggren, A., Nyman, M., Björck, I., Ahrne, S., Jeppsson, B. and Molin, G. (1998). Survival of Lactobacillus plantarum DSM 9843 (299v), and effect on the short-chain fatty acid content of faeces after ingestion of a rose-hip drink with fermented oats. International Journal of Food Microbiology, 42: 29–38.

Jounai, K., Ikado, K., Sugimura, T., Ano, Y., Braun, J. and Fujiwara, D. (2012). Spherical lactic acid bacteria activate plasmacytoid dendritic cells immunomodulatory function via TLR9-dependent crosstalk with myeloid dendritic cells. PloS One, 7: e32588.

Kailasapathy, K. and Chin, J. (2000). Survival and therapeutic potential of probiotic organisms with reference to Lactobacillus acidophilus and Bifidobacterium spp. Immunology and Cell Biology, 78: 80–88.

Kamiya, T., Wang, L., Forsythe, P., Goettsche, G., Mao, Y., Wang, Y., Tougas, G. and Bienenstock, J. (2006). Inhibitory effects of Lactobacillus reuteri on visceral pain induced by colorectal distension in Sprague-Dawley rats. Gut, 55: 191–196.

Karthik, L., Kumar, G., Keswani, T., Bhattacharyya, A., Chandar, S. S. and Bhaskara Rao, K. (2014). Protease inhibitors from marine actinobacteria as a potential source for antimalarial compound. PloS One, 9: e90972.

Khan, M. T., Nieuwdorp, M. and Bäckhed, F. (2014). Microbial modulation of insulin sensitivity. Cell Metabolism, 20: 753–760.

Khaneghah, A. M., Abhari, K., Eş, I., Soares, M. B., Oliveira, R. B., Hosseini, H., Rezaei, M., Balthazar, C. F., Silva, R. and Cruz, A. G. (2020). Interactions between probiotics and pathogenic microorganisms in hosts and foods: A review. Trends in Food Science & Technology, 95: 205–218.

Kim, S. -K., Guevarra, R. B., Kim, Y. -T., Kwon, J., Kim, H., Cho, J. H., Kim, H. B. and Lee, J. -H. (2019). Role of probiotics in human gut microbiome-associated diseases.

Krebs, B. (2016). Prebiotic and synbiotic treatment before colorectal surgery-randomised double blind trial. Collegium Antropologicum, 40: 35–40.

Kumamoto, Y. and Iwasaki, A. (2012). Unique features of antiviral immune system of the vaginal mucosa. Current Opinion in Immunology, 24: 411–416.

Kumar, P. S. (2013). Oral microbiota and systemic disease. Anaerobe, 24: 90–93.

Lamers, M. M., Beumer, J., Van Der Vaart, J., Knoops, K., Puschhof, J., Breugem, T. I., Ravelli, R. B., Paul van Schayck, J., Mykytyn, A. Z. and Duimel, H. Q. (2020). SARS-CoV-2 productively infects human gut enterocytes. Science, 369: 50–54.

Larsen, N., Vogensen, F. K., Gøbel, R., Michaelsen, K. F., Al-Soud, W. A., Sørensen, S. J., Hansen, L. H. and Jakobsen, M. (2011). Predominant genera of fecal microbiota in children with atopic dermatitis are not altered by intake of probiotic bacteria Lactobacillus acidophilus NCFM and Bifidobacterium animalis subsp. lactis Bi-07. FEMS Microbiology Ecology, 75: 482–496.

Lassenius, M. I., Pietiläinen, K. H., Kaartinen, K., Pussinen, P. J., Syrjänen, J., Forsblom, C., Pörsti, I., Rissanen, A., Kaprio, J. and Mustonen, J. (2011). Bacterial endotoxin activity in human serum is associated with dyslipidemia, insulin resistance, obesity, and chronic inflammation. Diabetes Care, 34: 1809–1815.

Lee, S. M., Donaldson, G. P., Mikulski, Z., Boyajian, S., Ley, K. and Mazmanian, S. K. (2013). Bacterial colonization factors control specificity and stability of the gut microbiota. Nature, 501: 426–429.

Lehtoranta, L., Latvala, S. and Lehtinen, M. J. (2020). Role of probiotics in stimulating the immune system in viral respiratory tract infections: A narrative review. Nutrients, 12: 3163.

Lepp, P. W., Brinig, M. M., Ouverney, C. C., Palm, K., Armitage, G. C. and Relman, D. A. (2004). Methanogenic Archaea and human periodontal disease. Proceedings of the National Academy of Sciences, 101: 6176–6181.

Ley, R. E., Peterson, D. A. and Gordon, J. I. (2006). Ecological and evolutionary forces shaping microbial diversity in the human intestine. Cell, 124: 837–848.

Li, H. and Cao, Y. (2010). Lactic acid bacterial cell factories for gamma-aminobutyric acid. Amino Acids, 39: 1107–1116.

Li, J., Sung, C. Y. J., Lee, N., Ni, Y., Pihlajamäki, J., Panagiotou, G. and El-Nezami, H. (2016). Probiotics modulated gut microbiota suppresses hepatocellular carcinoma growth in mice. Proceedings of the National Academy of Sciences, 113: E1306–E1315.

Li, L., Han, Z., Niu, X., Zhang, G., Jia, Y., Zhang, S. and He, C. (2019a). Probiotic supplementation for prevention of atopic dermatitis in infants and children: A systematic review and meta-analysis. American Journal of Clinical Dermatology, 20: 367–377.

Li, N., Ma, W. -T., Pang, M., Fan, Q. -L. and Hua, J. -L. (2019b). The commensal microbiota and viral infection: A comprehensive review. Frontiers in Immunology, 10: 1551.

Lieleg, O., Lieleg, C., Bloom, J., Buck, C. B. and Ribbeck, K. (2012). Mucin biopolymers as broad-spectrum antiviral agents. Biomacromolecules, 13: 1724–1732.

Lin, A., Bik, E. M., Costello, E. K., Dethlefsen, L., Haque, R., Relman, D. A. and Singh, U. (2013). Distinct distal gut microbiome diversity and composition in healthy children from Bangladesh and the United States. PLoS One, 8: e53838.

Liong, M. -T. (2008). Roles of probiotics and prebiotics in colon cancer prevention: Postulated mechanisms and *in vivo* evidence. International Journal of Molecular Sciences, 9: 854–863.

Lorea Baroja, M., Kirjavainen, P., Hekmat, S. and Reid, G. (2007). Anti-inflammatory effects of probiotic yogurt in inflammatory bowel disease patients. Clinical & Experimental Immunology, 149: 470–479.

Magalhaes, I., Pingris, K., Poitou, C., Bessoles, S., Venteclef, N., Kiaf, B., Beaudoin, L., Da Silva, J., Allatif, O. and Rossjohn, J. (2015). Mucosal-associated invariant T cell alterations in obese and type 2 diabetic patients. The Journal of Clinical Investigation, 125: 1752–1762.

Man, W. H., de Steenhuijsen Piters, W. A. and Bogaert, D. (2017). The microbiota of the respiratory tract: Gatekeeper to respiratory health. Nature Reviews Microbiology, 15: 259–270.

Marchesi, J. R. and Ravel, J. (2015). The vocabulary of microbiome research: A proposal. pp. 1–3, Springer.

Marteau, P., Seksik, P. and Jian, R. (2002). Probiotics and intestinal health effects: A clinical perspective. British Journal of Nutrition, 88: s51–s57.

Martín, V., Maldonado, A., Fernández, L., Rodríguez, J. M. and Connor, R. I. (2010). Inhibition of human immunodeficiency virus type 1 by lactic acid bacteria from human breastmilk. Breastfeeding Medicine, 5: 153–158.

Mayer, E. A. (2011). Gut feelings: The emerging biology of gut–brain communication. Nature Reviews Neuroscience, 12: 453–466.

Merenstein, D. and Salminen, S. (2017). Probiotics and prebiotics.

Miller, L. E., Lehtoranta, L. and Lehtinen, M. J. (2019). Short-term probiotic supplementation enhances cellular immune function in healthy elderly: Systematic review and meta-analysis of controlled studies. Nutrition Research, 64: 1–8.

Monaco, C. L., Gootenberg, D. B., Zhao, G., Handley, S. A., Ghebremichael, M. S., Lim, E. S., Lankowski, A., Baldridge, M. T., Wilen, C. B. and Flagg, M. (2016). Altered virome and bacterial microbiome in human immunodeficiency virus-associated acquired immunodeficiency syndrome. Cell Host & Microbe, 19: 311–322.

Monteiro-Sepulveda, M., Touch, S., Mendes-Sá, C., André, S., Poitou, C., Allatif, O., Cotillard, A., Fohrer-Ting, H., Hubert, E. -L. and Remark, R. (2015). Jejunal T cell inflammation in human obesity correlates with decreased enterocyte insulin signaling. Cell Metabolism, 22: 113–124.

Mueller, N. T., Bakacs, E., Combellick, J., Grigoryan, Z. and Dominguez-Bello, M. G. (2015). The infant microbiome development: Mom matters. Trends in Molecular Medicine, 21: 109–117.

Nair, M. S., Amalaradjou, M. and Venkitanarayanan, K. (2017). Antivirulence properties of probiotics in combating microbial pathogenesis. Advances in Applied Microbiology, 98: 1–29.

Nikonovas, T., Spessa, A., Doerr, S. H., Clay, G. D. and Mezbahuddin, S. (2020). Near-complete loss of fire-resistant primary tropical forest cover in Sumatra and Kalimantan. Communications Earth & Environment, 1: 1–8.

Nunn, K. L., Wang, Y. -Y., Harit, D., Humphrys, M. S., Ma, B., Cone, R., Ravel, J. and Lai, S. K. (2015). Enhanced trapping of HIV-1 by human cervicovaginal mucus is associated with Lactobacillus crispatus-dominant microbiota. MBio, 6: e01084–01015.

O'Toole, P., Marchesi, J. and Hill, C. (2017). Next-generation probiotics: the spectrum from probiotics to live biotherapeutics. Nat. Microbiol., 2(5): 17057.

Ochoa-Repáraz, J., Mielcarz, D. W., Ditrio, L. E., Burroughs, A. R., Foureau, D. M., Haque-Begum, S. and Kasper, L. H. (2009). Role of gut commensal microflora in the development of experimental autoimmune encephalomyelitis. The Journal of Immunology, 183: 6041–6050.

Ochoa-Repáraz, J., Mielcarz, D. W., Begum-Haque, S. and Kasper, L. H. (2011). Gut, bugs, and brain: Role of commensal bacteria in the control of central nervous system disease. Annals of Neurology, 69: 240–247.

Oh, B., Kim, B. -S., Kim, J. W., Kim, J. S., Koh, S. -J., Kim, B. G., Lee, K. L. and Chun, J. (2022). 幽门螺杆菌根除治疗过程中益生菌对肠道菌群的影响: 随机对照试验.

Olaya Galán, N., Ulloa Rubiano, J., Velez Reyes, F., Fernandez Duarte, K., Salas Cárdenas, S. and Gutierrez Fernandez, M. (2016). *In vitro* antiviral activity of Lactobacillus casei and Bifidobacterium adolescentis against rotavirus infection monitored by NSP 4 protein production. Journal of Applied Microbiology, 120: 1041–1051.

Ouwehand, A., Kirjavainen, P., Grönlund, M. -M., Isolauri, E. and Salminen, S. (1999). Adhesion of probiotic micro-organisms to intestinal mucus. International Dairy Journal, 9: 623–630.

Pace, N. R. (2009). Mapping the tree of life: progress and prospects. Microbiology and Molecular Biology Reviews, 73: 565–576.

Plaza-Diaz, J., Gomez-Llorente, C., Campaña-Martin, L., Matencio, E., Ortuño, I., Martínez-Silla, R., Gomez-Gallego, C., Periago, M. J., Ros, G. and Chenoll, E. (2013). Safety and immunomodulatory effects of three probiotic strains isolated from the feces of breast-fed infants in healthy adults: SETOPROB study. PloS One, 8: e78111.

Plovier, H., Everard, A., Druart, C., Depommier, C., Van Hul, M., Geurts, L., Chilloux, J., Ottman, N., Duparc, T. and Lichtenstein, L. (2017). A purified membrane protein from Akkermansia muciniphila or the pasteurized bacterium improves metabolism in obese and diabetic mice. Nature Medicine, 23: 107–113.

Qin, J., Li, R., Raes, J., Arumugam, M., Burgdorf, K. S., Manichanh, C., Nielsen, T., Pons, N., Levenez, F. and Yamada, T. (2010). A human gut microbial gene catalogue established by metagenomic sequencing. Nature, 464: 59–65.

Quintana, V. M., Torres, N. I., Wachsman, M. B., Sinko, P. J., Castilla, V. and Chikindas, M. (2014). Antiherpes simplex virus type 2 activity of the antimicrobial peptide subtilosin. Journal of Applied Microbiology, 117: 1253–1259.

Qureshi, N., Li, P. and Gu, Q. (2019). Probiotic therapy in Helicobacter pylori infection: A potential strategy against a serious pathogen? Applied Microbiology and Biotechnology, 103: 1573–1588.

Rhoades, N. S., Pinski, A. N., Monsibais, A. N., Jankeel, A., Doratt, B. M., Cinco, I. R., Ibraim, I. and Messaoudi, I. (2021). Acute SARS-CoV-2 infection is associated with an increased abundance of bacterial pathogens, including Pseudomonas aeruginosa in the nose. Cell Reports, 36: 109637.

Riiser, A. (2015). The human microbiome, asthma, and allergy. Allergy, Asthma & Clinical Immunology, 11: 1–7.

RodrÍGuez, J. J., Noristani, H. N., Hoover, W. B., Linley, S. B. and Vertes, R. P. (2011). Serotonergic projections and serotonin receptor expression in the reticular nucleus of the thalamus in the rat. Synapse, 65: 919–928.

Rogers, A. (2008). Molecular Oral Microbiology. Caister Academic Press.

Rutten, N., Gorissen, D., Eck, A., Niers, L., Vlieger, A., Besseling-Van Der Vaart, I., Budding, A., Savelkoul, P., Van der Ent, C. and Rijkers, G. (2015). Long term development of gut microbiota composition in atopic children: Impact of probiotics. PLoS One, 10: e0137681.

Salman, J. A. -S., Mahmood, N. N., Abdulsattar, B. O. and Abid, H. A. (2020). The effectiveness of probiotics against viral infections: a rapid review with focus on SARS-CoV-2 infection. Open Access Macedonian Journal of Medical Sciences, 8: 496–508.

Sanders, M. E., Heimbach, J. T., Pot, B., Tancredi, D. J., Lenoir-Wijnkoop, I., Lähteenmäki-Uutela, A., Gueimonde, M. and Bañares, S. (2011). Health Claims Substantiation for Probiotic and Prebiotic Products: Taylor & Francis.

Santos, F., Vera, J. L., van der Heijden, R., Valdez, G., de Vos, W. M., Sesma, F. and Hugenholtz, J. (2008). The complete coenzyme B12 biosynthesis gene cluster of Lactobacillus reuteri CRL1098. Microbiology, 154: 81–93.

Saulnier, D. M., Santos, F., Roos, S., Mistretta, T. -A., Spinler, J. K., Molenaar, D., Teusink, B. and Versalovic, J. (2011). Exploring metabolic pathway reconstruction and genome-wide expression profiling in Lactobacillus reuteri to define functional probiotic features. PLoS One, 6: e18783.

Schnadower, D., Tarr, P. I., Casper, T. C., Gorelick, M. H., Dean, J. M., O'Connell, K. J., Mahajan, P., Levine, A. C., Bhatt, S. R. and Roskind, C. G. (2018). Lactobacillus rhamnosus GG versus placebo for acute gastroenteritis in children. New England Journal of Medicine, 379: 2002–2014.

Schroeder, B. O. (2019). Fight them or feed them: How the intestinal mucus layer manages the gut microbiota. Gastroenterology Report, 7: 3–12.

Sender, R., Fuchs, S. and Milo, R. (2016). Revisiting the ratio of bacterial to host cells in humans. Cell, 337.

Shadnoush, M., Hosseini, R. S., Khalilnezhad, A., Navai, L., Goudarzi, H. and Vaezjalali, M. (2015). Effects of probiotics on gut microbiota in patients with inflammatory bowel disease: A double-blind, placebo-controlled clinical trial. Korean J. Gastroenterol., 65: 215–221.

Sharma, A., Lelic, D., Brock, C., Paine, P. and Aziz, Q. (2009). New technologies to investigate the brain-gut axis. World Journal of Gastroenterology: WJG, 15: 182.

Silverman, M. A., Konnikova, L. and Gerber, J. S. (2017). Impact of antibiotics on necrotizing enterocolitis and antibiotic-associated diarrhea. Gastroenterology Clinics, 46: 61–76.

Sokol, H., Pigneur, B., Watterlot, L., Lakhdari, O., Bermúdez-Humarán, L. G., Gratadoux, J. -J., Blugeon, S., Bridonneau, C., Furet, J. -P. and Corthier, G. (2008). Faecalibacterium prausnitzii is an anti-inflammatory commensal bacterium identified by gut microbiota analysis of Crohn disease patients. Proceedings of the National Academy of Sciences, 105: 16731–16736.

Sommer, F. and Bäckhed, F. (2013). The gut microbiota—masters of host development and physiology. Nature Reviews Microbiology, 11: 227–238.

Sonnenburg, J. L., Chen, C. T. L. and Gordon, J. I. (2006). Genomic and metabolic studies of the impact of probiotics on a model gut symbiont and host. PLoS Biology, 4: e413.

Su, M. S., Schlicht, S. and Gänzle, M. G. (2011). Contribution of glutamate decarboxylase in Lactobacillus reuteri to acid resistance and persistence in sourdough fermentation. pp. 1–12. Microbial Cell Factories: Springer.

Su, Y., Zhang, B. and Su, L. (2013). CD4 detected from Lactobacillus helps understand the interaction between Lactobacillus and HIV. Microbiological Research, 168: 273–277.

Szajewska, H. and Horvath, A. (2018). Lactobacillus rhamnosus GG in the primary prevention of eczema in children: A systematic review and meta-analysis. Nutrients, 10: 1319.

Szajewska, H., Horvath, A. and Piwowarczyk, A. (2010). Meta-analysis: The effects of Saccharomyces boulardii supplementation on Helicobacter pylori eradication rates and side effects during treatment. Alimentary Pharmacology & Therapeutics, 32: 1069–1079.

Szajewska, H. and Kołodziej, M. (2015). Systematic review with meta-analysis: Saccharomyces boulardii in the prevention of antibiotic-associated diarrhoea. Alimentary Pharmacology & Therapeutics, 42: 793–801.

Tabata, T., Petitt, M., Puerta-Guardo, H., Michlmayr, D., Wang, C., Fang-Hoover, J., Harris, E. and Pereira, L. (2016). Zika virus targets different primary human placental cells, suggesting two routes for vertical transmission. Cell Host & Microbe, 20: 155–166.

Tannock, G., Munro, K., Harmsen, H., Welling, G., Smart, J. and Gopal, P. (2000). Analysis of the fecal microflora of human subjects consuming a probiotic product containing Lactobacillus rhamnosus DR20. Applied and Environmental Microbiology, 66: 2578–2588.

Thomas, C. M., Hong, T., Van Pijkeren, J. P., Hemarajata, P., Trinh, D. V., Hu, W., Britton, R. A., Kalkum, M. and Versalovic, J. (2012). Histamine derived from probiotic Lactobacillus reuteri suppresses TNF via modulation of PKA and ERK signaling. PLoS One, 7: e31951.

Torres, N. I., Noll, K. S., Xu, S., Li, J., Huang, Q., Sinko, P.J., Wachsman, M. B., Chikindas and M. L. (2013). Safety, formulation and *in vitro* antiviral activity of the antimicrobial peptide subtilosin against herpes simplex virus type 1. Probiotics and Antimicrobial Proteins, 5: 26–35.

Triantafillidis, J. K., Nasioulas, G. and Kosmidis, P. A. (2009). Colorectal cancer and inflammatory bowel disease: Epidemiology, risk factors, mechanisms of carcinogenesis and prevention strategies. Anticancer Research, 29: 2727–2737.

van Zanten, G. C., Knudsen, A., Röytiö, H., Forssten, S., Lawther, M., Blennow, A., Lahtinen, S. J., Jakobsen, M., Svensson, B. and Jespersen, L. (2012). The effect of selected synbiotics on microbial composition and short-chain fatty acid production in a model system of the human colon.

Vanderhoof, J. A., Whitney, D. B., Antonson, D. L., Hanner, T. L., Lupo, J. V. and Young, R. J. (1999). Lactobacillus GG in the prevention of antibiotic-associated diarrhea in children. The Journal of Pediatrics, 135: 564–568.

Ventoulis, I., Sarmourli, T., Amoiridou, P., Mantzana, P., Exindari, M., Gioula, G. and Vyzantiadis, T. -A. (2020). Bloodstream infection by Saccharomyces cerevisiae in two COVID-19 patients after receiving supplementation of Saccharomyces in the ICU. Journal of Fungi, 6: 98.

Verdier, J., Luedde, T. and Sellge, G. (2015). Biliary mucosal barrier and microbiome. Visceral Medicine, 31: 156–161.

Versalovic, J. (2013). The human microbiome and probiotics: implications for pediatrics. Annals of Nutrition and Metabolism, 63: 42–52.

Vickers, N. J. (2017). Animal communication: When i'm calling you, will you answer too? Current Biology, 27: R713–R715.

Vrieze, A., Van Nood, E., Holleman, F., Salojärvi, J., Kootte, R. S., Bartelsman, J. F., Dallinga–Thie, G. M., Ackermans, M. T., Serlie, M. J. and Oozeer, R. (2012). Transfer of intestinal microbiota from lean donors increases insulin sensitivity in individuals with metabolic syndrome. Gastroenterology, 143: 913–916. e917.

Wachsman, M. B., Castilla, V., de Ruiz Holgado, A. P., de Torres, R. A., Sesma, F. and Coto, C. E. (2003). Enterocin CRL35 inhibits late stages of HSV-1 and HSV-2 replication *in vitro*. Antiviral Research, 58: 17–24.

Wall, R., Ross, R., Ryan, C., Hussey, S., Murphy, B., Fitzgerald, G. and Stanton, C. (2009). Role of gut microbiota in early infant development. Clinical Medicine. Pediatrics, 3:CMPed. S2008.

Wang, W., Xing, W., Wei, S., Gao, Q., Wei, X., Shi, L., Kong, Y. and Su, Z. (2019). Semi-rational screening of probiotics from the fecal flora of healthy adults against DSS-induced colitis mice by enhancing anti-inflammatory activity and modulating the gut microbiota.

Wang, Z., Yang, Y., Stefka, A., Sun, G. and Peng, L. (2014). Fungal microbiota and digestive diseases. Alimentary Pharmacology & Therapeutics, 39: 751–766.

Wassenaar, T. and Panigrahi, P. (2014). Is a foetus developing in a sterile environment? Letters in Applied Microbiology, 59: 572–579.

Wells, J. M., Brummer, R. J., Derrien, M., MacDonald, T. T., Troost, F., Cani, P. D., Theodorou, V., Dekker, J., Méheust, A. and De Vos, W. M. (2017). Homeostasis of the gut barrier and potential biomarkers. American Journal of Physiology-Gastrointestinal and Liver Physiology, 312: G171–G193.

Wen, H., Yin, X., Yuan, Z., Wang, X. and Su, S. (2018). Comparative analysis of gut microbial communities in children under 5 years old with diarrhea.

Wikoff, W. R., Anfora, A. T., Liu, J., Schultz, P. G., Lesley, S. A., Peters, E. C. and Siuzdak, G. (2009). Metabolomics analysis reveals large effects of gut microflora on mammalian blood metabolites. Proceedings of the National Academy of Sciences, 106: 3698–3703.

Willey, J. M., Sherwood, L. and Woolverton, C. J. (2011). Prescott's Microbiology. McGraw-Hill New York.

Wu, G. D., Chen, J., Hoffmann, C., Bittinger, K., Chen, Y. -Y., Keilbaugh, S. A., Bewtra, M., Knights, D., Walters, W. A. and Knight, R. (2011). Linking long-term dietary patterns with gut microbial enterotypes. Science, 334: 105–108.

Wu, Y., Guo, C., Tang, L., Hong, Z., Zhou, J., Dong, X., Yin, H., Xiao, Q., Tang, Y. and Qu, X. (2020). Prolonged presence of SARS-CoV-2 viral RNA in faecal samples. The Lancet Gastroenterology & Hepatology, 5: 434–435.

Yang, A. T. M. I. T., Phillips, A. and Tian, Y. -C. (2007). Original article.

Yang, I., Corwin, E. J., Brennan, P. A., Jordan, S., Murphy, J. R. and Dunlop, A. (2016). The infant microbiome: Implications for infant health and neurocognitive development. Nursing Research, 65: 76.

Yang, J. and Yu, J. (2018). The association of diet, gut microbiota and colorectal cancer: What we eat may imply what we get. Protein & Cell, 9: 474–487.

Yokote, H., Miyake, S., Croxford, J. L., Oki, S., Mizusawa, H. and Yamamura, T. (2008). NKT cell-dependent amelioration of a mouse model of multiple sclerosis by altering gut flora. The American Journal of Pathology, 173: 1714–1723.

Zabihollahi, R., Motevaseli, E., Sadat, S. M., Azizi-Saraji, A. R., Asaadi-Dalaie, S. and Modarressi M. H. (2012). Inhibition of HIV and HSV infection by vaginal lactobacilli *in vitro* and *in vivo*. DARU Journal of Pharmaceutical Sciences, 20: 1–7.

Zhang, C., Derrien, M., Levenez, F., Brazeilles, R., Ballal, S. A., Kim, J., Degivry, M. -C., Quéré, G., Garault, P. and van Hylckama Vlieg, J. E. (2016). Ecological robustness of the gut microbiota in response to ingestion of transient food-borne microbes. The ISME Journal, 10: 2235–2245.

Zhao, M., Shen, C. and Ma, L. (2018). Treatment efficacy of probiotics on atopic dermatitis, zooming in on infants: A systematic review and meta-analysis. International Journal of Dermatology, 57: 635–641.

Zhernakova, A., Kurilshikov, A., Bonder, M. J., Tigchelaar, E. F., Schirmer, M., Vatanen, T., Mujagic, Z., Vila, A. V., Falony, G. and Vieira-Silva, S. (2016). Population-based metagenomics analysis reveals markers for gut microbiome composition and diversity. Science, 352: 565–569.

Zhou, Y., Gao, H., Mihindukulasuriya, K. A., La Rosa, P. S., Wylie, K. M., Vishnivetskaya, T., Podar, M., Warner, B., Tarr, P. I. and Nelson, D. E. (2013). Biogeography of the ecosystems of the healthy human body. Genome Biology, 14: 1–18.

Zmora, N., Zilberman-Schapira, G., Suez, J., Mor, U., Dori-Bachash, M., Bashiardes, S., Kotler, E., Zur, M., Regev-Lehavi, D. and Brik, R. B. -Z. (2018). Personalized gut mucosal colonization resistance to empiric probiotics is associated with unique host and microbiome features. Cell, 174: 1388–1405. e1321.

CHAPTER 8

The Effect of Microbiome Exchange on Humans and Animals

Karma G. Dolma,[1,*] *Shanmuga Sundar S.*,[2] *Chamma Gupta*[3] and *Veeranoot Nissapatorn*[4,5]

Introduction

The interaction of the microbial environment is often overlooked in a larger sense. Human contact with animals occurs directly through livestock, consuming animal products, keeping pets and the presence of wildlife. Dealing with animals is an essential element of existence in the environment. With advancements made for the survival of mankind, in the form of encroachment and disturbances in the anthropogenic aspects, the potential for the transmission of microflora is omnipresent. As we see

[1] Department of Microbiology, Sikkim Manipal Institute of Medical Sciences, Sikkim Manipal University, Sikkim, India.
[2] Department of Biotechnology, Aarupadai Veedu Institute of Technology, Vinayak Mission's Research Foundation (DU), Paiyanoor-603104, Chennai, Tamil Nadu, India.
[3] Department of Biotechnology, Sikkim Manipal Institute of Medical Sciences, Sikkim Manipal University, Sikkim, India.
[4] Department of Medical Technology, School of Allied Health Sciences, Walailak University, Nakhon Si Thammarat, 80160, Thailand.
[5] School of Allied Health Sciences, Southeast Asia water Team (SEA Water Team) and World Union for Herbal Drug Discovery (WUHeDD), and Research Excellence center for Innovation and Health Products, Walailak University, Nakhon Si Thammarat, 80160, Thailand.
* Corresponding author: kgdolma@outlook.com

an increasing interest in the role of the microbiome in human biology, we seldom disregard the relation between human and animal exposure. This can not only lead to possible transmission of diseases, but there may also be simultaneous exposure and corresponding transmission of the commensal flora of the animals. There is a surge to divulge more about human-animal interactions among different microbial populations. This may be found in different human populations or races, thereby shaping the fundamental basis of the fact that different microbiomes may be associated with a particular human population. Awareness regarding these transmission routes has always reflected on the pathogenic microbes and had a hand in zoonotic transmission. Along with them, we know that commensal microbes are also transmitted through these routes, albeit at a lower rate.

Human Microbiome

In the human gut, the presence of certain bacteria dwelling and networking with the human body is defined as the human microbiota. (Grice and Segre 2011). The human microbiota is made up of 10–100 trillion symbiotic microbial cells anchored by an individual, predominantly bacteria in the gut (Turnbaugh et al. 2007). The term 'microbiota' (microbial taxa pertaining to humans) and 'microbiome' (microbial list and their genes) are often used concurrently (Ursell et al. 2012). The term 'human microbiome' refers to the genetic make-up of microorganisms (microbiota) that reside at a certain locus within the human body. Several functional and structural regions, including the skin, mucosa, gastrointestinal system, respiratory tract, urogenital tract, etc., are inhabited by microorganisms. The ecology they create is intricate and distinct, and it can adjust to the specific environmental requirements of each niche (Whiteside et al. 2015). The microbiome of humans is incessantly altering in adaptation to the host. The human microbiome is mostly determined by factors like age, dietary habits, lifestyle, fluctuating hormones, inherited genes and underlying diseases (Figures 1 and 2). Dysbiosis, a transformation in the conformation of the human microbiota, can, however, end in serious disorders. A healthy microbiome has been found to be decisive for preserving and upholding good health (Turnbaugh et al. 2007). A large proportion of the microbiome is localised in the gastrointestinal tract. The primary elements accountable for conserving and nourishing human health are these organisms. A dysbiotic flora in the intestine can perhaps be related to changes in the immunological atmosphere, as shown by earlier investigations on the human microbiome project. Dysbiosis has also been connected to other potentially fatal medical diseases, such as cancer, cardiovascular diseases, bowel inflammation and bacterial infections, that are challenging to manage because of antibiotic resistance (Morgan and Huttenhower 2012, Pascal et al. 2018).

It turns out that the majority of the microbes in our bodies were acquired from our mothers – via the birth canal, skin-to-skin contact, and lactation. The process of becoming exposed to these primordial germs is known as seeding, and it is affected by the surroundings in which one is born. The microbial mosaic is influenced by an array of factors, including nature, nutrition, family, pets, etc. The microbial flora attains a stable state of microbiome in a couple of years. For the remaining period of our lives, external influences (including nutrition, exercise, medication and sleep) continue to alter our body composition. According to estimates, there are approximately 10^{13}–10^{14} microbial cells in the human microbiota, for every human cell. These figures are based on the amount of bacteria present in the colon (3.8×10^{13} bacteria), the organ that houses the majority of microbes (Sender et al. 2016). In the flora of the gut microbiota, there are basically three main phyla of bacteria known more commonly: Firmicutes, Bacteroidetes and Actinobacteria (Tap et al. 2009). The varied and intricate microbiome is generally considered a well-designed part of the human genome, where it has at least 50 to 100-fold more genes compared to the human host. These have an added feature where we can observe more than one type of non-coded protein performing an important role in initiating metabolism, thus aiding in the functioning of human physiology (Hooper and Gordon 2001).

Factors Affecting Human Gut

The signal of many molecules are transmitted by the gut microbiota that modulates the structure of the surface and regulates its composition. The intestinal epithelial cells (IEC) consisting of immunoglobulin A (IgA), antimicrobial peptides (AMPs) and mucus produce several molecules that encourage the growth of some microbial spp., and can also inhibit the growth of others. Mucin is known for harmonizing the shape of the gut microbiota. In the large intestine, mucins have an important role in the formation of the host microflora. In the lower intestine, there are no indications of any microorganism, whereas the outer layer of the intestine has soluble O-glycans mucins. It is known to provide a nutrient source by initiation or a building site for the microbiota of the gut, thereby ensuring that the microbes are away from the intestinal epithelial cells (Artis et al. 2004). Mucus and mucin O-glycans, along with the help of glycoside hydrolases and polysaccharide lyases, help in choosing the most appropriate microbial species of the host (Tailford et al. 2015). In terms of the small intestine, where the AMPs play a key role in selecting and shaping the gut microbiota, it is to be known that the mucus level is lower in the small intestine. In the microbe-associated molecular patterns (MAMP) system, the structural components like lipopolysaccharide and flagella stimulate the pattern recognition receptor (PRR) (Hooper 2009). The PRR system through the structural components and metabolites, like the paneth cell, induce the production of AMPs in the gut. The PRR-MAMP assists in the function of mucin barriers and prompts the activation of IgA, mucin glycoproteins, and AMPs (Carvalho et al. 2012). AMPs are proteins that include defensins, cathelicidins, Reg family proteins, and ribonucleases (Cash et al. 2006), produced by intestinal epithelial cells. These proteins act as the body's first line of defence against intruders. They are known to directly kill various microorganisms, as well as cancerous cells. Studies have demonstrated that the most prevalent or dominant organisms of the gut microbiota can resist the high concentration of AMPs, especially when the host distinguishes commensals from pathogenic bacteria (Cullen et al. 2015).

The IEC promotes the growth of some bacterial species and inhibits the growth of others by secreting certain secretions. In this way, the host can control the shape and structure of its gut microbiota by moderating its secretor. The presence of dephosphorylate lipid A (Raetz and Whitfield 2002) in Bacteroides, one of the most common gram negative bacteria in the gut flora, has the ability to resist the high concentration of AMPs in the intestine (Cullen et al. 2015). Hexameric pore formation in gram positive bacterial membranes moulded by lectins is antibacterial in nature, thereby averting them from reaching the intestinal mucus layer (Mukherjee et al. 2014). Thus, the secretors and the IEC help the host control the shape and structure of the gut microflora. The secretory immunoglobulin A (SIgA) produced by plasma cells in the intestinal mucosa has certain features, like controlling the bacterial load (Macpherson and Uhr 2004) as well as aiding in the formation of biofilm by binding the SIgA receptors to bacteria (Bollinger et al. 2003). The micro-RNAs (miRNAs) are produced in the nucleus and are transferred to the cytoplasm (Djuranovic et al. 2012). They are non-coding RNA and are usually about 18–23 nucleotides in length. miRNAs are specific host factors that can also be used to assess the shape and structure of the gut microbiota. They are known to exist outside the cells and can very well circulate in body fluids (Weber et al. 2010). Moreover, one miRNA can efficiently target different mRNAs (Kalla et al. 2015). Several studies (Ahmed et al. 2009, Link et al. 2012, Liu et al. 2016) have demonstrated the role of miRNAs in the gut microbiota composition, through the investigation of stool and intestinal content. Researchers (Link et al. 2012) have studied the relationship between the deficiency of IEC, miRNA and gut dysbiosis. The mechanism associated with wild-type faecal transplantation that may restore the gut microbiota

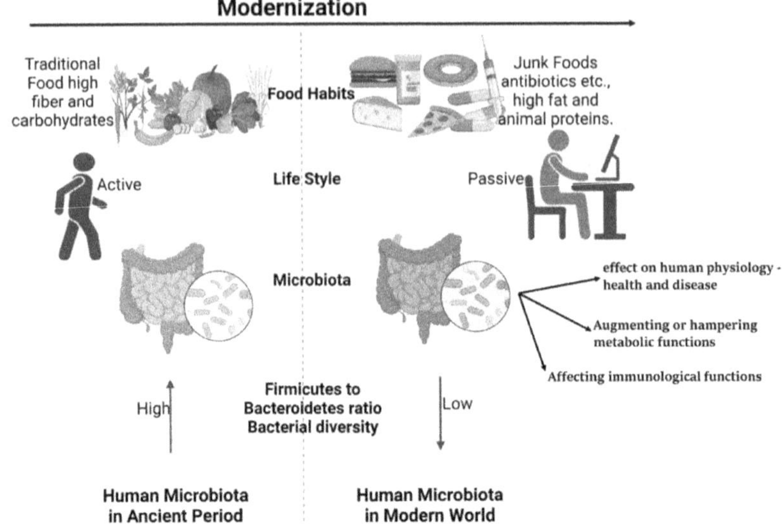

Figure 1. Impact of Modernization on Microbiota Composition and Diversity. It illustrates the dynamic changes in microbial communities in response to various aspects of modernization. It demonstrates the shift in microbiota composition between traditional and modern lifestyles. The traditional lifestyle, represented on the left, is characterized by a diverse and balanced microbial community. In contrast, the modern lifestyle, depicted on the right, shows alterations in microbiota composition, with potential reductions in diversity.

has also been presented, indicating that miRNAs are able to regulate the gut microbiome. Several miRNAs can then enter the gut bacterial cells and regulate their growth and gene expression.

Factors Influencing Homeostasis of Microbiota

The gut microbiota is established early in life, but can be altered later by various factors that affect its development and diversity. Numerous studies have reported the presence of several bacterial products in the meconium (Jiménez et al. 2008, Nagpal et al. 2017, Wampach et al. 2017), the amniotic fluid and the placenta (Friedrich 2013). This indicates that the initiation of colonisation of the intestinal environment by the microbiota is known to begin in the placenta or the amniotic fluid. The study where the oral administration of *Enterococcus fecium* in pregnant mice resulted in the isolation of the said organisms in the stool samples of the newborn due to the interchange between the digestive tract and amniotic fluid via the blood stream (Jiménez et al. 2008). Further, it has been confirmed with evidence that maternal microbes can be found in the amniotic fluid as well as the placenta (Goldenberg et al. 2000). The mechanism of delivery of the newborn is also known to affect the development of the gut microbiota. The presence of Lactobacillus and Prevotella has been predominantly associated with vaginally delivered cases, while caesarean delivery cases have a gut microbiota with the presence of Streptococcus, Corynebacterium and Propionibacterium (Dominguez-Bello et al. 2010, Mackie et al. 1999). Over time, we observe that these primary organisms of the microbiota evolve and become more diverse and relatively stable; after a period of 2–3 years, their flora is more similar to the adult microbiota (Becker-Dreps et al. 2015). The composition of the primary microbiota acquired during the initial period serves as the foundation of community; hence, it is very important for a newborn to have a good microflora (Groer et al. 2014). Hence, the findings of several studies indicate that breastfed babies have a more stable population of good mucosal immune responses (Bezirtzoglou et al. 2011). The diet of the newborn is also one of the primary factors shaping the early gut microbiota. This is where the role of different types of milk comes into play and further elucidates the outcome. Breast milk is basically oligosaccharides, which can be easily broken down by Lactobacillus and Bifidobacterium, leading to an increase in the manifestation of immunoglobulin G (Ouwehand et al. 2002). Other dominant species like Enterococcus, Enterbacteria, Bacteroides, Clostiridia and Streptococcus are found to be more invested in formula-fed newborns. Several immunological activities are modulated by the milk microbiota. These are known to increase the number of plasma cells to produce more IgA (Gross 2007), stimulate cytotoxic Th1 cells, develop the local and systemic immune systems (Houghteling and Walker 2015), and activate and produce cytokines, NK cells, CD4+ T-cells and CD8+ T-cells by certain strains of Lactobacillus. Variation in one's diet is also known to influence the development and diversity of the gut microbiota. Dominant bacteria like Ruminococcus, Roseburia, and Eubacterium are found especially in the vegetarian diet and are known to metabolise insoluble carbohydrates. In the case of a non-vegetarian diet, there is a decrease in Firmicutes and an increase in Bacteroides. Evidence of the fermentation of amino acids, the production of short-chain fatty acids as an energy source, and the formation of harmful end products has also been observed in a non-vegetarian diet (Windey et al. 2012).

The introduction of antibiotics is known for both harmful and beneficial actions. It can eliminate the pathological as well as the beneficial flora of the gut indiscriminately (Table 1). The competitive exclusion mechanism through which the elimination of the pathological microorganism is carried out can be uniformly disturbed by the use of antibiotics, simultaneously promoting the growth of other pathogenic organisms like *Clostridium difficile* (Ramnani et al. 2012).

Table 1. Usage of certain antibiotics causing the disruption of the gut flora.

Antibiotics	Role	References
Clindamycin	Deviations in the microbiota flora; absence of Bacteroides diversity	(Jernberg et al. 2007)
Clarithromycin	Minimisation in the population of Actinobacteria	(Jakobsson et al. 2010)
Ciprofloxacin	Decrease in Ruminococcus population	(Dethlefsen and Relman 2011)
Vancomycin	Alterations in the gut microbiota; presence of recurrent C. difficile infection; encouraged growth of pathological strains of *E. coli*	(Zar et al. 2007)
Vancomycin	Diminution of Bacteroidetes, Fuminococcus and Faecalibacterium flora; increase of Proteobacteria species	(Dethlefsen and Relman 2011)

Usage of Certain Antibiosis Causing the Disruption of the Gut Flora

The use of antibiotics has intensified in agriculture, particularly in the farming of poultry and beef. The antibiotics are routinely given to the livestock to increase their growth and weight. The presence of antibiotics in food has a terrible impact on human health. Further, this is supplemented by the use of pesticides and other harmful chemicals that are sprayed on food and other consumables. However, there is a lack of evidence of its harmful effects on the gut microbiota (Lee et al. 2017).

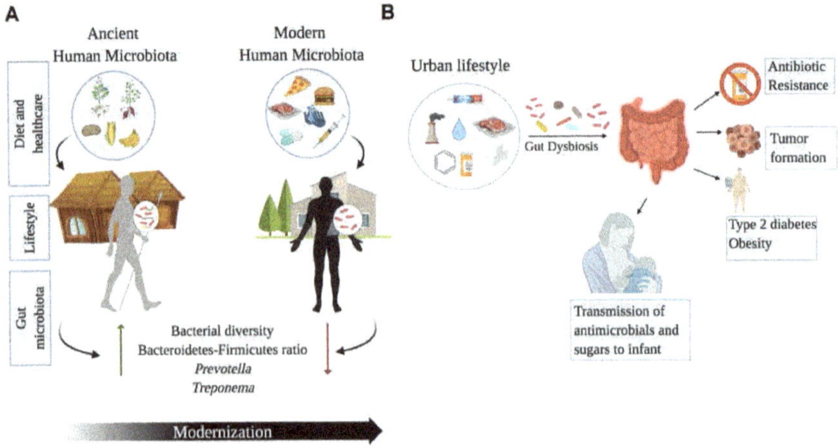

Figure 2. Illustrates the complex interplay between modernization, urbanization, and the increased prevalence of various illnesses. It depicts the transition from traditional to modern urban lifestyles, showcasing changes in living environments, dietary habits, and sedentary behaviours.

Animal Microbiome

The microbiome exhibited by several non-human vertebrate species, including livestock and wildlife species (e.g., Tasmanian devil, red panda, giant panda, black howler monkey and koala), has been sequenced in the last few years (Bahrndorff et al. 2016). Insects have colonised a wide variety of habitats and are one of the most diverse and populous animal groups on the planet. Therefore, it is not surprising that many different insect species are habitats for huge and varied microbial ecosystems that are crucial to insect biology (May 1988). A wide range of different organisms live inside different insect species; in particular, microbial populations in the digestive tract have received much scrutiny (Weiss and Aksoy 2011). Some insect species have a substantially more diversified microbiome when viewed alongside other insect species. In contrast with the microbes of other species, such as fruit flies or mosquitoes, certain synanthropic insects, such as the green bottle fly, have microbiomes that tend to be far more varied (Chandler et al. 2011). The abundance of species may be a result of synanthropic flies living and reproducing in beddings made of decomposing organic matter rich in microorganisms and/or animal manure. There are microbiomes existing in different sets of invertebrates, albeit for a smaller number of species. Five families of marine invertebrates, among others, have been the subject of studies comparing the microbiomes of various species of marine invertebrates with or without photosynthetic symbionts (Bourne et al. 2013). Marine species (e.g., oysters) of commercial value have also been addressed in literature (Lokmer and Mathias Wegner 2015), and invertebrate soil microbes have drawn some interest as well. The intestinal microbes of soil animals are crucial for food digestion and play a crucial ecological role in the global carbon cycle. According to recent investigations, a few soil invertebrates, including collembolans, earthworms and nematodes, have a rich flora of microbial communities and potential symbionts, similar to that of terrestrial insects (Ladygina et al. 2009). Additionally, findings have demonstrated how variations in feeding among earthworm ecological niches trigger the emergence of various bacterial populations (Thakuria et al. 2010). Moreover, changing the soil ecosystem may have an effect on the ecology and functioning of earthworms by affecting the bacterial community associated with their gut walls. Although the ecosystem of microbes in invertebrates such as collembolans and earthworms remains to be thoroughly investigated, there exists compelling proof that gastrointestinal microbes may contribute to the deterioration of recalcitrant biological compounds like chitin and lignocellulose (Egert et al. 2004).

Factors Affecting the Animal Microbiome

Multifaceted forms of association of the host with its microbiome may be present in animals, including symbiotic and pathogenic associations (Weiss and Aksoy 2011). The hosts may benefit from symbiotic microbial communities in a number of ways, including nutritional augmentation, host immune system, and social connections (Ridley et al. 2012). The existence of a core microbiota is implied by the fact that the gut symbionts of many insects are essential for their survival and growth (Hosokawa et al. 2006). Animal-microbial connections tend to be adaptable and facultative, and the host may harbour various symbionts during various periods. Symbionts need not be completely reliant on the host (Weiss and Aksoy 2011). Numerous biotic and abiotic variables may additionally have an impact on the relationship between the host and the microbial community. These elements may impact the host's immune system, dietary habits, reproduction, interaction and numerous other systems (Lam et al. 2007). Findings on the influence of the microbiota on the well-being of animals are few and virtually exclusively confined to human subjects. Nevertheless, quite a few investigations have examined the impact of individual bacterial symbionts on animal health, with a focus on insect types in particular (Feldhaar 2011). The stakes are increasingly high for comprehending what elements can change an animal's microbiota, as a means to determine how health is impacted and to elucidate the variations among ecosystems, species, and/or communities. The colonies of bacteria in invertebrate and vertebrate animals appear to be influenced by a variety of variables like the host's genetic make-up, eating habits, lifespan phase, lab training, and the environmental and physiological state of, for instance, the animal' gut (Wong et al. 2013). Additionally, current evidence suggests that the microbiome affects the immune system's stability and/or growth, resistance to external changes, and intake of nutrients (Weiss and Aksoy 2011). Several invertebrates do not have varied and intricate relationships with microbes. Studies that seek to comprehend the role of certain bacteria and the overall microbiota in host physiological activities are made possible by such insect model systems. For instance, *Drosophila melanogaster* offers a good model system to investigate a few of the aforementioned issues, and it is feasible to raise axenic flies of this species. The functional effects of particular bacterial populations and the host's complete microbiome can be thoroughly examined using next-generation sequencing techniques. Results of studies on *D. melanogaster* have demonstrated how the microbiota alters its metabolic activity and the allocation of carbohydrates in a lab setting (Ridley et al. 2012). Comparable functional investigations of the ant microbiota indicates a high potential for cellulose degradation (Suen et al. 2010), while the metabolic processes of microorganisms in herbivorous species play an essential part in fixing, recycling or reforming nitrogen (Anderson et al. 2012). Additionally, the idea has been put forth that variations in the gut microbiota may contribute to variations in herbivorous insect and their capacity to ingest chemically protected plants (Hammer and Bowers 2015).

Certain features (like external fertilisation, optical transparency of embryos and larvae, an adaptive immune system, and the presence of systemic features like the liver, intestines, etc.) make the zebrafish a simple yet indispensable model where complex host-microbiota interactions can be understood. In a study, the dominant constituents of the microbiota included ϒ-Proteobacteria and Fusobacteria classes (Roeselers et al. 2011). In the zebrafish intestinal environment, the host anatomy, physiology, nutrient availability and immunology initiate the selective pressure for the presence of highly specific microbes, rather than the environment or diet (Kostic et al. 2013). A study on the dietary lipid absorption of the microbiota in the transparent zebrafish embryo demonstrated an increase in fatty acid absorption in conventional zebrafish compared with germ-free fish, under both fed and starved dietary conditions. Earlier reports have also demonstrated that a high-calorie intake can influence the relative abundance of Firmicutes and Bacteroidetes (Ley et al. 2008, Semova et al. 2012).

At the host's genetic level, about 99% of mouse genes are similar to humans and undoubtedly share key similarities with the human gut microbiome. However, certain mouse traits, like coprophagia, the structure of the oesophagus and the GIT system, skin, etc., are significantly different from those of humans and influence the microbial communities. Nucleotide Oligomerization Domain (NOD) 2 mutations are a risk factor for human Crohn's disease. NOD-2 is expressed by the gut microbiome. NOD-2 knocks out mouse pathogens, and mice deficient in NOD-1 and 2 have an altered microbiota composition, compared to other animals (Petnicki-Ocwieja et al. 2009).

Antibiotics also effectively disrupt the gut microbiota. A study has highlighted how Vancomycin Resistant Enterococcus (VRE) can dominate the gut microbiome and expose it to VRE bacteraemia (Ubeda et al. 2012). Another study investigated as to how long antibiotics promote growth in livestock and humans; therein, it was also concluded that exposure to antibiotics at an early age can render a child susceptible to obesity (Cho et al. 2012).

Conclusion

There is a necessity to quantify the host phenotypes by means of anthropometrics and survey data across several more inhabitants to gauge if there is any augmented microbial assortment or specific microbial conformation(s) associated with increased microbial operation and if these result in constructive or negative host health consequences. Without analysing microbial function in transitional populations as well, we can fail to contemplate the health discrepancies associated with being exposed to a wide range of microbes.

The representation of understudied populations in future studies is central in order to understand the diversity in the pattern of microbiomes across the human species and to precisely identify the large disparity of human-animal interaction effects on the human microbiome. Different exchanges with animals contribute towards shaping the human microbiome across populations. Microbial transmission at human-animal boundaries is of serious concern for emerging zoonotic pathogens. Hence, understanding dissimilarities in human microbial revelations across inhabitants is vital for moderating disease risk, especially in developing countries.

References

Ahmed, F. E., Jeffries, C. D., Vos, P. W., Flake, G., Nuovo, G. J., Sinar, D. R., Naziri, W. and Marcuard, S. P. (2009). Diagnostic microRNA markers for screening sporadic human colon cancer and active ulcerative colitis in stool and tissue. Cancer Genomics & Proteomics, 6(5): 281–295.

Anderson, K. E., Russell, J. A., Moreau, C. S., Kautz, S., Sullam, K. E., Hu, Y., Basinger, U., Mott, B. M., Buck, N. and Wheeler, D. E. (2012). Highly similar microbial communities are shared among related and trophically similar ant species: Trophically similar gut communities. Molecular Ecology, 21(9): 2282–2296. https://doi.org/10.1111/j.1365-294X.2011.05464.x.

Artis, D., Wang, M. L., Keilbaugh, S. A., He, W., Brenes, M., Swain, G. P., Knight, P. A., Donaldson, D. D., Lazar, M. A., Miller, H. R. P., Schad, G. A., Scott, P. and Wu, G. D. (2004). RELMbeta/FIZZ2 is a goblet cell-specific immune-effector molecule in the gastrointestinal tract. Proceedings of the National Academy of Sciences of the United States of America, 101(37): 13596–13600. https://doi.org/10.1073/pnas.0404034101.

Bahrndorff, S., Alemu, T., Alemneh, T. and Lund Nielsen, J. (2016). The microbiome of animals: implications for conservation biology. International Journal of Genomics, 2016: e5304028. https://doi.org/10.1155/2016/5304028.

Becker-Dreps, S., Allali, I., Monteagudo, A., Vilchez, S., Hudgens, M. G., Rogawski, E. T., Carroll, I. M., Zambrana, L. E., Espinoza, F. and Andrea Azcarate-Peril, M. (2015). Gut microbiome composition in young nicaraguan children during diarrhea episodes and recovery. The American Journal of Tropical Medicine and Hygiene, 93(6): 1187–1193. https://doi.org/10.4269/ajtmh.15-0322.

Bezirtzoglou, E., Tsiotsias, A. and Welling, G. W. (2011). Microbiota profile in feces of breast- and formula-fed newborns by using fluorescence in situ hybridization (FISH). Anaerobe, 17(6): 478–482. https://doi.org/10.1016/j.anaerobe.2011.03.009.

Bollinger, R. R., Everett, M. L., Palestrant, D., Love, S. D., Lin, S. S. and Parker, W. (2003). Human secretory immunoglobulin A may contribute to biofilm formation in the gut. Immunology, 109(4): 580–587. https://doi.org/10.1046/j.1365-2567.2003.01700.x.

Bourne, D. G., Dennis, P. G., Uthicke, S., Soo, R. M., Tyson, G. W. and Webster, N. (2013). Coral reef invertebrate microbiomes correlate with the presence of photosymbionts. The ISME Journal, 7(7): Article 7. https://doi.org/10.1038/ismej.2012.172.

Carvalho, F. A., Aitken, J. D., Vijay-Kumar, M. and Gewirtz, A. T. (2012). Toll-like receptor-gut microbiota interactions: Perturb at your own risk! Annual Review of Physiology, 74: 177–198. https://doi.org/10.1146/annurev-physiol-020911-153330.

Cash, H. L., Whitham, C. V., Behrendt, C. L. and Hooper, L. V. (2006). Symbiotic bacteria direct expression of an intestinal bactericidal lectin. Science (New York, N.Y.), 313(5790): 1126–1130. https://doi.org/10.1126/science.1127119.

Chandler, J. A., Morgan Lang, J., Bhatnagar, S., Eisen, J. A. and Kopp, A. (2011). Bacterial communities of diverse drosophila species: Ecological context of a host–microbe model system. PLoS Genetics, 7(9): e1002272. https://doi.org/10.1371/journal.pgen.1002272.

Cho, I., Yamanishi, S., Cox, L., Methé, B. A., Zavadil, J., Li, K., Gao, Z., Mahana, D., Raju, K., Teitler, I., Li, H., Alekseyenko, A. V. and Blaser, M. J. (2012). Antibiotics in early life alter the murine colonic microbiome and adiposity. Nature, 488(7413): 621–626. https://doi.org/10.1038/nature11400.

Cullen, T. W., Schofield, W. B., Barry, N. A., Putnam, E. E., Rundell, E. A., Trent, M. S., Degnan, P. H., Booth, C. J., Yu, H. and Goodman, A. L. (2015). Antimicrobial peptide resistance mediates resilience of prominent gut commensals during inflammation. Science (New York, N.Y.), 347(6218): 170–175. https://doi.org/10.1126/science.1260580.

Dethlefsen, L. and Relman, D. A. (2011). Incomplete recovery and individualized responses of the human distal gut microbiota to repeated antibiotic perturbation. Proceedings of the National Academy of Sciences of the United States of America, 108 Suppl 1(Suppl 1): 4554–4561. https://doi.org/10.1073/pnas.1000087107.

Djuranovic, S., Nahvi, A. and Green, R. (2012). MiRNA-mediated gene silencing by translational repression followed by mRNA deadenylation and decay. Science (New York, N.Y.), 336(6078): 237–240. https://doi.org/10.1126/science.1215691.

Dominguez-Bello, M. G., Costello, E. K., Contreras, M., Magris, M., Hidalgo, G., Fierer, N. and Knight, R. (2010). Delivery mode shapes the acquisition and structure of the initial microbiota across multiple body habitats in newborns. Proceedings of the National Academy of Sciences of the United States of America, 107(26): 11971–11975. https://doi.org/10.1073/pnas.1002601107.

Egert, M., Marhan, S., Wagner, B., Scheu, S. and Friedrich, M. W. (2004). Molecular profiling of 16S rRNA genes reveals diet-related differences of microbial communities in soil, gut, and casts of Lumbricus terrestris L. (Oligochaeta: Lumbricidae). FEMS Microbiology Ecology, 48(2): 187–197. https://doi.org/10.1016/j.femsec.2004.01.007.

Feldhaar, H. (2011). Bacterial symbionts as mediators of ecologically important traits of insect hosts. Ecological Entomology, 36(5): 533–543. https://doi.org/10.1111/j.1365-2311.2011.01318.x.

Friedrich, M. J. (2013). Genomes of microbes inhabiting the body offer clues to human health and disease. JAMA, 309(14): 1447–1449. https://doi.org/10.1001/jama.2013.2824.

Goldenberg, R. L., Hauth, J. C. and Andrews, W. W. (2000). Intrauterine infection and preterm delivery. The New England Journal of Medicine, 342(20): 1500–1507. https://doi.org/10.1056/NEJM200005183422007.

Grice, E. A. and Segre, J. A. (2011). The skin microbiome. Nature Reviews. Microbiology, 9(4): 244–253. https://doi.org/10.1038/nrmicro2537.

Groer, M. W., Luciano, A. A., Dishaw, L. J., Ashmeade, T. L., Miller, E. and Gilbert, J. A. (2014). Development of the preterm infant gut microbiome: A research priority. Microbiome, 2: 38. https://doi.org/10.1186/2049-2618-2-38.

Gross, L. (2007). Microbes colonize a baby's gut with distinction. PLoS Biology, 5(7): e191. https://doi.org/10.1371/journal.pbio.0050191.

Hammer, T. J. and Bowers, M. D. (2015). Gut microbes may facilitate insect herbivory of chemically defended plants. Oecologia, 179(1): 1–14.

Hooper, L. V. (2009). Do symbiotic bacteria subvert host immunity? Nature reviews. Microbiology, 7(5): 367–374. https://doi.org/10.1038/nrmicro2114.

Hooper, L. V. and Gordon, J. I. (2001). Commensal host-bacterial relationships in the gut. Science (New York, N.Y.), 292(5519): 1115–1118. https://doi.org/10.1126/science.1058709.

Hosokawa, T., Kikuchi, Y., Nikoh, N., Shimada, M. and Fukatsu, T. (2006). Strict host-symbiont cospeciation and reductive genome evolution in insect gut bacteria. PLoS Biology, 4(10): e337. https://doi.org/10.1371/journal.pbio.0040337.

Houghteling, P. D. and Walker, W. A. (2015). Why is initial bacterial colonization of the intestine important to the infant's and child's health? Journal of Pediatric Gastroenterology and Nutrition, 60(3): 294–307. https://doi.org/10.1097/MPG.0000000000000597.

Jakobsson, H. E., Jernberg, C., Andersson, A. F., Sjölund-Karlsson, M., Jansson, J. K. and Engstrand, L. (2010). Short-term antibiotic treatment has differing long-term impacts on the human throat and gut microbiome. PLOS ONE, 5(3): e9836. https://doi.org/10.1371/journal.pone.0009836.

Jernberg, C., Löfmark, S., Edlund, C. and Jansson, J. K. (2007). Long-term ecological impacts of antibiotic administration on the human intestinal microbiota. The ISME Journal, 1(1): 56–66. https://doi.org/10.1038/ismej.2007.3.

Jiménez, E., Marín, M. L., Martín, R., Odriozola, J. M., Olivares, M., Xaus, J., Fernández, L. and Rodríguez, J. M. (2008). Is meconium from healthy newborns actually sterile? Research in Microbiology, 159(3): 187–193. https://doi.org/10.1016/j.resmic.2007.12.007.

Kalla, R., Ventham, N. T., Kennedy, N. A., Quintana, J. F., Nimmo, E. R., Buck, A. H. and Satsangi, J. (2015). MicroRNAs: New players in IBD. Gut, 64(3): 504–517. https://doi.org/10.1136/gutjnl-2014-307891.

Kostic, A. D., Howitt, M. R. and Garrett, W. S. (2013). Exploring host-microbiota interactions in animal models and humans. Genes & Development, 27(7): 701–718. https://doi.org/10.1101/gad.212522.112.

Ladygina, N., Johansson, T., CanbÃ¤ck, B., Tunlid, A. and Hedlund, K. (2009). Diversity of bacteria associated with grassland soil nematodes of different feeding groups: Bacteria associated with grassland soil nematodes. FEMS Microbiology Ecology, 69(1): 53–61. https://doi.org/10.1111/j.1574-6941.2009.00687.x.

Lam, K., Babor, D., Duthie, B., Babor, E. -M., Moore, M. and Gries, G. (2007). Proliferating bacterial symbionts on house fly eggs affect oviposition behaviour of adult flies. Animal Behaviour, 74(1): 81–92. https://doi.org/10.1016/j.anbehav.2006.11.013.

Lee, Y. -M., Kim, K. -S., Jacobs, D. R. and Lee, D. -H. (2017). Persistent organic pollutants in adipose tissue should be considered in obesity research: Obesity and persistent organic pollutants. Obesity Reviews, 18(2): 129–139. https://doi.org/10.1111/obr.12481.

Ley, R. E., Hamady, M., Lozupone, C., Turnbaugh, P. J., Ramey, R. R., Bircher, J. S., Schlegel, M. L., Tucker, T. A., Schrenzel, M. D., Knight, R. and Gordon, J. I. (2008). Evolution of mammals and their gut microbes. Science (New York, N.Y.), 320(5883): 1647–1651. https://doi.org/10.1126/science.1155725.

Link, A., Becker, V., Goel, A., Wex, T. and Malfertheiner, P. (2012). Feasibility of fecal microRNAs as novel biomarkers for pancreatic cancer. PLoS ONE, 7(8): e42933. https://doi.org/10.1371/journal.pone.0042933.

Liu, S., da Cunha, A. P., Rezende, R. M., Cialic, R., Wei, Z., Bry, L., Comstock, L. E., Gandhi, R. and Weiner, H. L. (2016). The host shapes the gut microbiota via fecal microRNA. Cell Host & Microbe, 19(1): 32–43. https://doi.org/10.1016/j.chom.2015.12.005.

Lokmer, A. and Mathias Wegner, K. (2015). Hemolymph microbiome of Pacific oysters in response to temperature, temperature stress and infection. The ISME Journal, 9(3): Article 3. https://doi.org/10.1038/ismej.2014.160.

Mackie, R. I., Sghir, A. and Gaskins, H. R. (1999). Developmental microbial ecology of the neonatal gastrointestinal tract. The American Journal of Clinical Nutrition, 69(5): 1035S–1045S. https://doi.org/10.1093/ajcn/69.5.1035s.

Macpherson, A. J. and Uhr, T. (2004). Induction of protective IgA by intestinal dendritic cells carrying commensal bacteria. Science (New York, N.Y.), 303(5664): 1662–1665. https://doi.org/10.1126/science.1091334.

May, R. M. (1988). How many species are there on Earth? Science (New York, N.Y.), 241(4872): 1441–1449. https://doi.org/10.1126/science.241.4872.1441.

Morgan, X. C. and Huttenhower, C. (2012). Chapter 12: Human microbiome analysis. PLoS Computational Biology, 8(12): e1002808. https://doi.org/10.1371/journal.pcbi.1002808.

Mukherjee, S., Zheng, H., Derebe, M., Callenberg, K., Partch, C. L., Rollins, D., Propheter, D. C., Rizo, J., Grabe, M., Jiang, Q. -X. and Hooper, L. V. (2014). Antibacterial membrane attack by a pore-forming intestinal C-type lectin. Nature, 505(7481): 103–107. https://doi.org/10.1038/nature12729.

Nagpal, R., Tsuji, H., Takahashi, T., Nomoto, K., Kawashima, K., Nagata, S. and Yamashiro, Y. (2017). Ontogenesis of the gut microbiota composition in healthy, full-term, vaginally born and breast-fed infants over the first 3 years of life: A quantitative bird's-eye view. Frontiers in Microbiology, 8: 1388. https://doi.org/10.3389/fmicb.2017.01388.

Ouwehand, A., Isolauri, E. and Salminen, S. (2002). The role of the intestinal microflora for the development of the immune system in early childhood. European Journal of Nutrition, 41 Suppl 1: I32–37. https://doi.org/10.1007/s00394-002-1105-4.

Pascal, M., Perez-Gordo, M., Caballero, T., Escribese, M. M., Lopez Longo, M. N., Luengo, O., Manso, L., Matheu, V., Seoane, E., Zamorano, M., Labrador, M. and Mayorga, C. (2018). Microbiome and allergic diseases. Frontiers in Immunology, 9: 1584. https://doi.org/10.3389/fimmu.2018.01584.

Petnicki-Ocwieja, T., Hrncir, T., Liu, Y. -J., Biswas, A., Hudcovic, T., Tlaskalova-Hogenova, H. and Kobayashi, K. S. (2009). Nod2 is required for the regulation of commensal microbiota in the intestine. Proceedings of the National Academy of Sciences of the United States of America, 106(37): 15813–15818. https://doi.org/10.1073/pnas.0907722106.

Raetz, C. R. H. and Whitfield, C. (2002). Lipopolysaccharide endotoxins. Annual Review of Biochemistry, 71: 635–700. https://doi.org/10.1146/annurev.biochem.71.110601.135414.

Ramnani, P., Chitarrari, R., Tuohy, K., Grant, J., Hotchkiss, S., Philp, K., Campbell, R., Gill, C. and Rowland, I. (2012). *In vitro* fermentation and prebiotic potential of novel low molecular weight polysaccharides derived from agar and alginate seaweeds. Anaerobe, 18(1): 1–6. https://doi.org/10.1016/j.anaerobe.2011.08.003.

Ridley, E. V., Wong, A. C. -N., Westmiller, S. and Douglas, A. E. (2012). Impact of the resident microbiota on the nutritional phenotype of drosophila melanogaster. PLOS ONE, 7(5): e36765. https://doi.org/10.1371/journal.pone.0036765.

Roeselers, G., Mittge, E. K., Stephens, W. Z., Parichy, D. M., Cavanaugh, C. M., Guillemin, K. and Rawls, J. F. (2011). Evidence for a core gut microbiota in the zebrafish. The ISME Journal, 5(10): 1595–1608. https://doi.org/10.1038/ismej.2011.38.

Semova, I., Carten, J. D., Stombaugh, J., Mackey, L. C., Knight, R., Farber, S. A. and Rawls, J. F. (2012). Microbiota regulate intestinal absorption and metabolism of fatty acids in the zebrafish. Cell Host & Microbe, 12(3): 277–288. https://doi.org/10.1016/j.chom.2012.08.003.

Sender, R., Fuchs, S. and Milo, R. (2016). Revised estimates for the number of human and bacteria cells in the body. PLoS Biology, 14(8): e1002533. https://doi.org/10.1371/journal.pbio.1002533.

Suen, G., Scott, J. J., Aylward, F. O., Adams, S. M., Tringe, S. G., Pinto-Tomás, A. A., Foster, C. E., Pauly, M., Weimer, P. J., Barry, K. W., Goodwin, L. A., Bouffard, P., Li, L., Osterberger, J., Harkins, T. T., Slater, S. C., Donohue, T. J. and Currie, C. R. (2010). An insect herbivore microbiome with high plant biomass-degrading capacity. PLoS Genetics, 6(9): e1001129..

Tailford, L. E., Crost, E. H., Kavanaugh, D. and Juge, N. (2015). Mucin glycan foraging in the human gut microbiome. Frontiers in Genetics, 6: 81. https://doi.org/10.3389/fgene.2015.00081.

Tap, J., Mondot, S., Levenez, F., Pelletier, E., Caron, C., Furet, J. -P., Ugarte, E., Muñoz-Tamayo, R., Paslier, D. L. E., Nalin, R., Dore, J. and Leclerc, M. (2009). Towards the human intestinal microbiota phylogenetic core. Environmental Microbiology, 11(10): 2574–2584. https://doi.org/10.1111/j.1462-2920.2009.01982.x.

Thakuria, D., Schmidt, O., Finan, D., Egan, D. and Doohan, F. M. (2010). Gut wall bacteria of earthworms: A natural selection process. The ISME Journal, 4(3): Article 3. https://doi.org/10.1038/ismej.2009.124.

Turnbaugh, P. J., Ley, R. E., Hamady, M., Fraser-Liggett, C. M., Knight, R. and Gordon, J. I. (2007). The human microbiome project. Nature, 449(7164): 804–810. https://doi.org/10.1038/nature06244.

Ubeda, C., Lipuma, L., Gobourne, A., Viale, A., Leiner, I., Equinda, M., Khanin, R. and Pamer, E. G. (2012). Familial transmission rather than defective innate immunity shapes the distinct intestinal microbiota of TLR-deficient mice. The Journal of Experimental Medicine, 209(8): 1445–1456. https://doi.org/10.1084/jem.20120504.

Ursell, L. K., Metcalf, J. L., Parfrey, L. W. and Knight, R. (2012). Defining the human microbiome. Nutrition Reviews, 70 Suppl 1(Suppl 1): S38–44. https://doi.org/10.1111/j.1753-4887.2012.00493.x.

Wampach, L., Heintz-Buschart, A., Hogan, A., Muller, E. E. L., Narayanasamy, S., Laczny, C. C., Hugerth, L. W., Bindl, L., Bottu, J., Andersson, A. F., de Beaufort, C. and Wilmes, P. (2017). Colonization and succession within the human gut microbiome by archaea, bacteria, and microeukaryotes during the first year of life. Frontiers in Microbiology, 8: 738. https://doi.org/10.3389/fmicb.2017.00738.

Weber, J. A., Baxter, D. H., Zhang, S., Huang, D. Y., Huang, K. H., Lee, M. J., Galas, D. J. and Wang, K. (2010). The microRNA spectrum in 12 body fluids. Clinical Chemistry, 56(11): 1733–1741. https://doi.org/10.1373/clinchem.2010.147405.

Weiss, B. and Aksoy, S. (2011). Microbiome influences on insect host vector competence. Trends in Parasitology, 27(11): 514–522. https://doi.org/10.1016/j.pt.2011.05.001.

Whiteside, S. A., Razvi, H., Dave, S., Reid, G. and Burton, J. P. (2015). The microbiome of the urinary tract—A role beyond infection. Nature Reviews. Urology, 12(2): 81–90. https://doi.org/10.1038/nrurol.2014.361.

Windey, K., De Preter, V. and Verbeke, K. (2012). Relevance of protein fermentation to gut health. Molecular Nutrition & Food Research, 56(1): 184–196. https://doi.org/10.1002/mnfr.201100542.

Wong, A. C. -N., Chaston, J. M. andDouglas, A. E. (2013). The inconstant gut microbiota of Drosophila species revealed by 16S rRNA gene analysis. The ISME Journal, 7(10): Article 10. https://doi.org/10.1038/ismej.2013.86.

Zar, F. A., Bakkanagari, S. R., Moorthi, K. M. L. S. T. andDavis, M. B. (2007). A comparison of vancomycin and metronidazole for the treatment of Clostridium difficile-associated diarrhea, stratified by disease severity. Clinical Infectious Diseases: An Official Publication of the Infectious Diseases Society of America, 45(3): 302–307. https://doi.org/10.1086/519265.

CHAPTER 9

Unravelling Gut Microbiomes, Viromes, and Biofilm Investigations in the Developing Next-generation Sequencing Era

Bhagwan Narayan Rekadwad,[1,*]
Narayan Dattatraya Totewad,[2] *Nanditha Pramod*[1]
and *Mangesh V. Suryavanshi*[3]

Introduction

The study of microorganisms and microbiomes has a long history, dating back to the 1670s when Anton van Leeuwenhoek first observed microorganisms using a microscope. In the late 19th century, Robert Koch and Louis Pasteur established the Germ Theory of Disease, proposing that microorganisms are responsible for many human and animal diseases (National Research Council (US) 2004, Slotten 2014). Thus, researchers embarked on investigations into the variety and distribution of microorganisms found in myriad environments, including water and soil.

[1] Division of Microbiology and Biotechnology, Yenepoya Research Centre, Yenepoya (Deemed to be University), Mangalore - 575018, Karnataka, India.
[2] Department of Microbiology, B. K. Birla College of Arts, Science & Commerce (Autonomous), Kalyan - 421 304, Thane, Maharashtra, India.
[3] Cardiovascular and Metabolic Sciences Department, Lerner Research Institute, Cleveland Clinic, OH 44195, United States.
* Corresponding author: rekadwad@gmail.com

The identification and characterization of a huge number of previously unknown microorganisms and the study of their functions in a variety of ecological processes has resulted in vast research in the context of microbiomes, viromes and biofilms (Table 1) – in the subject of Microbiology (Opal 2009). This has led to the concept of the microbial ecosystem, and the realization that microorganisms play important roles in many ecological processes. In the late 20th century, advances in DNA sequencing technology enabled scientists to study the diversity and composition of microbial communities in greater detail (Prosser et al. 2007). This fueled the development of the field of metagenomics – the study of genetic material from microbial communities. This field has greatly expanded our understanding of the diversity and functions of microorganisms in different environments, including the human body (Handelsman 2004). In recent years, the study of the human gut microbiome has attracted significant attention due to the growing knowledge of the gut microbiome's significance in human health and disease (Martin et al. 2014). An increased understanding of the role of the gut microbiome in digestion, nutrient absorption, immune system development, and production of certain vitamins and amino acids (Rowland et al. 2018) is necessary to avoid health related issues and reap biotechnological benefits. In the era of next-generation sequencing technology and bioinformatics, investigations have demonstrated progress in the subject of microorganisms and microbiomes – from the discovery and isolation of individual microorganisms to the study of their communities and functions within various environments, human, animals and plants (Gilbert et al. 2018). This chapter describes the progression of gut microbiomes, viromes and biofilm investigations in the developing next-generation sequencing era, from the concept of microbial cell to the future of microbiome after a century.

Microbiome – Diet, Health and Disease

Gut Microbiome

The term 'microbiome' refers to the collection of all the microorganisms that live in a particular environment, such as the human gut or the soil. These microorganisms mainly include bacteria, archaea, fungi, algae, lichens and viruses. Additionally, protozoa may be added to answer the question: *"What are the different kinds of microorganisms?"* A separate subject or term has already been coined for studies on the virus community known as the 'virome' (Berg et al. 2020). These groups of prokaryotic and eukaryotic microorganisms are extremely important to human health as

Table 1. Important events and discoveries on microbiomes, viromes and biofilms since 1990s.

Microbiome	
Identification of the first human gut microbial species, *Methanobrevibacter smithii*	Miller et al. 1982
Discovery of the first example of horizontal gene transfer between bacteria in the human gut microbiome	Dethlefsen et al. 2007
Demonstration of a causal link between gut microbiota dysbiosis and inflammatory bowel disease	Sartor 2008
Identification of a gut microbial signature associated with obesity and metabolic syndrome	Turnbaugh et al. 2009
Discovery of a bacteriophage that can kill antibiotic-resistant bacteria in the human gut microbiome	Reyes et al. 2010
Demonstration of the role of the vaginal microbiome in maintaining women's reproductive health	Ravel et al. 2011
Identification of a gut microbial signature associated with autism spectrum disorder	Wang et al. 2011
Characterization of the oral microbiome and its role in oral health and disease	Dewhirst et al. 2013
Identification of a gut microbial signature associated with Parkinson's disease	Sampson et al. 2016
Demonstration of the role of the gut microbiome in regulating the immune system and influencing susceptibility to infection	Belkaid and Hand 2014
Discovery of a gut microbial pathway that influences cardiovascular disease risk	Tang et al. 2017
Identification of a microbial metabolite that may help prevent age-related cognitive decline	Kundu et al. 2021
Virome	
Discovery of the first bacteriophage, the lytic phage, in the early 20th century	d'Herelle 1917
Identification of bacteriophages as a major component of the human microbiome	Minot et al. 2013
Characterization of the gut virome and its role in regulating the bacterial community and host immune system	Reyes et al. 2010
Discovery of a virophage, a virus that infects other viruses, in the ocean	La Scola et al. 2008
Demonstration of the role of bacteriophages in shaping bacterial evolution and ecology	Suttle 2005
Identification of the gut virome as a potential source of novel antimicrobial agents	Dalmasso et al. 2014

Table 1 contd. ...

...Table 1 contd.

Discovery of a virus that can target and kill cancer cells, opening up new avenues for cancer therapy	Russell et al. 2012
Characterization of the virome associated with the respiratory tract and its role in respiratory infections	Jung et al. 2017
Demonstration of the role of the virome in modulating the gut-brain axis and influencing behavior and mood	Yolken and Severance 2015
Identification of a novel family of giant viruses, the Klosneuviruses, in an underground aquifer	Schulz et al. 2017
Discovery of a virus that can infect bacteria of the human gut microbiome and modify their behavior	Cornuault et al. 2018
Biofilm	
First observation of bacterial biofilms on medical implants and their impact on device-associated infections	Christensen et al. 1985
Demonstration of the role of quorum sensing in regulating biofilm formation and dispersion	Davies et al. 1998
Identification of the extracellular polymeric substances (EPS) matrix as a critical component of biofilm structure and function	Flemming and Wingender 2010
Characterization of the metabolic diversity and spatial organization of microbial communities within biofilms	Stewart and Franklin 2008
Demonstration of the adaptive evolution of bacteria within biofilms and their increased resistance to antibiotics and host immune responses	Mah and O'Toole 2001
Identification of the role of biofilms in persistent infections, including chronic wounds and infections associated with cystic fibrosis	Costerton et al. 1999
Development of novel strategies for disrupting biofilm formation and dispersal, including quorum sensing inhibitors and EPS-degrading enzymes	Flemming et al. 2016
Discovery of the role of biofilms in biocorrosion and fouling of industrial systems, leading to the development of new strategies for biofilm control	Flemming et al. 2016
Identification of interactions between biofilms and the host immune system, and the development of new immunotherapeutic approaches for biofilm-associated infections	Bjarnsholt et al. 2013
Demonstration of the potential of biofilms for bioremediation and bioprocessing applications, including wastewater treatment and bioenergy production	Flemming et al. 2016
Characterization of the role of biofilms in the ecology and evolution of microbial communities, including their impact on nutrient cycling and biogeochemical processes	Renslow et al. 2018

they are components of microbiomes, viromes and biofilms. They have genes that help them grow and that affect or help the way in which important parts of the human body work, like the digestive system, the immune system, and the central nervous system (Grice and Segre 2012, Willis and Gabaldón 2020, Matijašić et al. 2020). The composition of an individual's microbiome is entirely unique to them, and it might fluctuate over time in response to age, food habits, sleep pattern, travel patterns, intake of water source, medication, and the presence of infectious agents (Conlon and Bird 2014, Leeming et al. 2019, Gibiino et al. 2021, Begum et al. 2022).

Various health conditions, such as inflammatory bowel disease, obesity and certain autoimmune disorders, have been associated with dysbiosis, which is a disruption in the microbiome (Gupta et al. 2016). A segment of the microbial community secretes extra-cellular products that provide additional structural support to build the host immune system, e.g., calcium chelation (Limoli et al. 2015, Repac Antić et al. 2022), bile acids for treatment of intestinal inflammation (Godlewska et al. 2022), and cGAS-STING-IFN-I axis system for promoting host resistance (Erttmann et al. 2022). The analysis of the gut microbiome has shown that the gut microbiota can influence nutrient absorption, metabolism, and the development of chronic diseases such as obesity (Breton et al. 2022, Barone et al. 2022), diabetes and gut leakiness (Ye et al. 2022) and inflammatory bowel disease (Pisani et al. 2022), which may be aided by changes in the dietary habit of the individual. Different diet combinations (such as high or low fat, high- or low-fiber diets, high or low water intake, spicy or less spicy food eating habits, etc.) have distinct effects on the composition and diversity of the gut microbiome. For example, a diet high in fiber can promote the growth of beneficial bacteria such as *Bifidobacterium* and *Prevotella* (Tomova et al. 2019, Cronin et al. 2021), while a diet high in fat and sugar can lead to an overgrowth of harmful bacteria such as Firmicutes (Murphy et al. 2015, Moughaizel et al. 2022).

The gut microbiome has a profound effect on immune function as well. The gut microbiome helps in training the immune system and promoting immune tolerance, and an imbalance in the gut microbiome can lead to chronic inflammation and an increased risk of autoimmune diseases (Wang et al. 2022, Mousa et al. 2022, Hou et al. 2022). As discussed earlier, research has shown that the gut microbiome can influence nutrient absorption, metabolism, and the development of chronic diseases such as obesity, diabetes and inflammatory bowel disease. An imbalance in the gut microbiome, known as dysbiosis, is believed to be a significant risk factor for these and other chronic health conditions (Lin et al. 2022,

Zhang 2022). Moreover, the gut microbiome may also play a role in the development of certain types of cancer, such as colon cancer. Studies have shown that certain gut bacteria produce carcinogenic compounds, and an imbalance in the gut microbiome may promote the growth of cancer cells (Sadrekarimi et al. 2022, Siddiqui et al. 2022, Al-Ishaq et al. 2022). More recent studies have also suggested a link between gut microbiome and mental health (such as depression and anxiety).

The gut-brain axis is the communication pathway between the gut and the brain, and it has been found that gut microbiome plays a crucial role in this communication (Radjabzadeh et al. 2022, Vera-Urbina et al. 2022, Zhu et al. 2022). Hence, the gut microbiome is an essential component of human health and well-being, and maintaining a diverse and balanced gut microbiome is crucial for reducing the risk of chronic diseases (Gomma 2020, Hou et al. 2022). Eating a healthy and balanced diet rich in fruits, vegetables, and whole grains, and avoiding processed foods is a great way to support the health of the gut microbiome (Leeuwendaal et al. 2022).

Ways for Microbiome Exchange – Benefits and Infection

The infection cycle of a microorganism is a complex process that involves a variety of mechanisms that enable it to establish an infection in a host, replicate and spread, evade the host's immune response, and persist inside the host. These pathways include adherence, invasion, replication, evasion, dissemination, persistence and chronic infection. Adherence refers to attaching to a surface or tissue in the host; invasion involves invading the host tissue; replication includes increasing in numbers and spreading to new host tissues; evasion involves hiding in host cells; and dissemination refers to infecting new tissues and organs. Persistence deals with remaining in the host for long periods of time without causing symptoms (Thakur et al. 2019). The best example to justify the above is *Mycobacterium tuberculosis* (Chandra et al. 2022). The relationships between community microbial species are termed as microbial co-occurrence. It can reveal functional relationships, identify keystone species and explain microbial community structure and function. Microbial interactions determine microbial community composition and function, and networks can visualize and analyze these interactions (Gao et al. 2022b).

For example, genus *Bifidobacterium* is known for their ability to ferment carbohydrates and produce lactic acid, which can help maintain a healthy pH balance in the gut and support the growth of other beneficial bacteria. They also play a role in the production of certain vitamins, such as Vitamin K and B12. For Vitamin K, *Bifidobacterium* uses Menaquinone biosynthesis pathway which starts with the conversion of Phylloquinone

(Vitamin K1) to menadione (MK-4), through the action of menaquinone 7-epoxidase. The pathways for the synthesis of Vitamin B12 are not fully understood; however, there are some proposed mechanisms. For Vitamin B12, *Bifidobacterium* uses the Corrinoid B12 biosynthesis pathway which starts with the conversion of uroporphyrinogen III to precorrin-2, through the action of uroporphyrinogen III methyltransferase (Dailey et al. 2017, Indira et al. 2019, Balabanova et al. 2021, Watane et al. 2022, Kato et al. 2022). Human gut bacteria include *Prevotella,* that affects human health and causes diseases. *Prevotella* overgrowth in the gut microbiome increases the risk of mucosal inflammation. TNF-alpha and IL-6 levels in the gut are elevated by some *Prevotella* species. Inflammatory disorders, including IBD and colitis, are linked to *Prevotella* overgrowth. Prevotella and other bacteria may play a role in mucosal inflammation as well (Iljazovic et al. 2021). *Prevotella* helps the gut digest complex carbohydrates like dietary fibers and resistant starches. This contrasts with Western populations, wherein it increases inflammation and metabolic disorders. *Prevotella* species can efficiently break down complex carbohydrates and produce gut-healthy short-chain fatty acids (SCFAs) (Nogal et al. 2021).

Gut Virome

The gut virome is an important component of the human microbiome and has significant implications for human health and disease (Zheng et al. 2020). The gut virome is highly diverse and is influenced by factors such as diet, antibiotics and the host immune system. Hence, the gut virome plays vital roles in various aspects of human health, including immune function, metabolism and neurological function (Spencer et al. 2022). The gut virome is also implicated in the development of certain diseases, such as inflammatory bowel disease (IBD), colorectal cancer and type 1 diabetes. Moreover, there are a few challenges associated with studying the gut virome, including the complexity of viral communities and the lack of standardized methods for virome analysis (Fulci et al. 2021). This suggests that further research is needed to better understand the gut virome and its role in human health and disease. Thus, the gut virome is a relatively new component of the human microbiome, and a deeper understanding of the gut virome may lead to new therapeutic and diagnostic strategies for a range of human diseases (Cao et al. 2022). The gut virome is thought to play an important role in regulating the gut microbiome by controlling the abundance and diversity of bacterial populations (Mukhopadhya et al. 2019, Tiamani et al. 2022).

Viruses depend on the genetic machinery of host cells, which is diverse and difficult to understand and functions under different circumstances. Human microbial ecosystems are complex due to the dynamic interactions of bacteriophage viruses with their cellular hosts (Sutton and Hill 2019). Human viral populations have been studied on a large scale as a result of advances in sequencing and bioinformatics tools. Human viral community variety is connected to the diversity of their cellular hosts, fast evolution, horizontal gene transfers, and intimate interactions with host nucleic acids (Goettsch et al. 2021, Irwin et al. 2022). Virus genotypes vary in different environments – oral, gastric, respiratory, and bloodstream, which were previously thought to be sterile. When mucosal defenses are impaired, virome members can cross mucosal barriers. Human viral communities can carry genes with important functions in host disease, including antibiotic resistance. Close contacts share oral virobiota, which could spread antibiotic resistance to healthy people (Alaoui Mdarhri et al. 2022, Campos-Madueno et al. 2023). Viruses can affect ecosystem dynamics by transferring beneficial gene functions or lysing certain populations of cellular hosts (i.e., viruses act as regulators) (Gao et al. 2022a), which can improve microbial resilience to disturbances, immune evasion, physiological processes, and pathogen colonization (Abeles and Pride 2014, Dong et al. 2022). Viruses that have helped shape our evolution by becoming regulators involve endogenization (the incorporation of viral DNA into the genome of the host, leading to the formation of new genes or modification of existing genes) (Blyton et al. 2022, Richert-Pöggeler et al. 2022).

Horizontal gene transfer (HGT) is the transfer of genetic material between different microbial species (Lean et al. 2022) that affect the microbiome and have implications on human health and disease conditions (Borodovich et al. 2022). Speciation is the process by which viruses can aid in the formation of new species by erecting genetic barriers that prevent interbreeding between different populations, thereby increasing the number of different species on Earth (Radman 2022, Lal et al. 2022). Adaptation is the occurrence of new genetic variations due to natural selection, that are beneficial for the host and help the host acclimatize to new environments and changing conditions (LaTourrette and Garcia-Ruiz 2022, He et al. 2022). Population bottlenecks are caused by viral outbreaks and have led to the reduction of genetic diversity within a population, posing both negative and positive effects on the population's ability to adapt to new challenges (Somovilla et al. 2022, Carvalho et al. 2023). The gut virome has been shown to have the potential to directly influence human health, according to research. Some viral infections, for instance, have been linked to inflammatory bowel disease as well as other gut disorders (Mukhopadhya et al. 2019, Tiamani et al. 2022). Additionally, it is possible that the gut virome is involved in the regulation of host immunity, which may have repercussions for the development of cancer and other chronic diseases (Ungaro et al. 2019). In spite of this, it is abundantly clear that the gut virome plays a crucial role in human health, animal health, and disease. Therefore, when researching the gut ecosystem, it is essential to take into account both the gut microbiome and the gut virome.

Microbial Biofilms – Good, Bad and Worst

Microbial biofilms have good, bad as well as the worst aspects. The good thing about a microbial biofilm is that it is a complex population of bacteria adhered to a surface and encased in extracellular polymeric molecules (EPS). This cycle becomes irreversible once the biofilm matures and attaches to the matrix on various substrata (Figure 1). Bacteria, fungus and algae can create biofilms on rocks, soil, water and medical devices. Biofilms are a diverse colony of microorganisms that can benefit or harm an ecosystem. Understanding the roles of biofilms in varied ecosystems helps control and maximize their benefits while reducing their drawbacks. The bad about microbial biofilms is that they can both benefit or harm ecosystems. Biofilms can break down contaminants, provide habitats, and cycle nutrients. Biofilms in water treatment plants remove contaminants. But biofilms can also cause diseases and reduce industrial efficiency. They can also build on medical devices like catheters and industrial equipment, causing corrosion and infections. Microbial biofilms can form microscopic layers on teeth, in the urinary tract, and at other places, causing chronic and hard-to-treat illnesses. Biofilms in the gut microbiota can cause persistent infections. Infections caused by microbial biofilms are bad. Antimicrobial-resistant bacteria biofilms are hard to eradicate. Biofilms on catheters can again induce chronic, hard-to-treat illnesses.

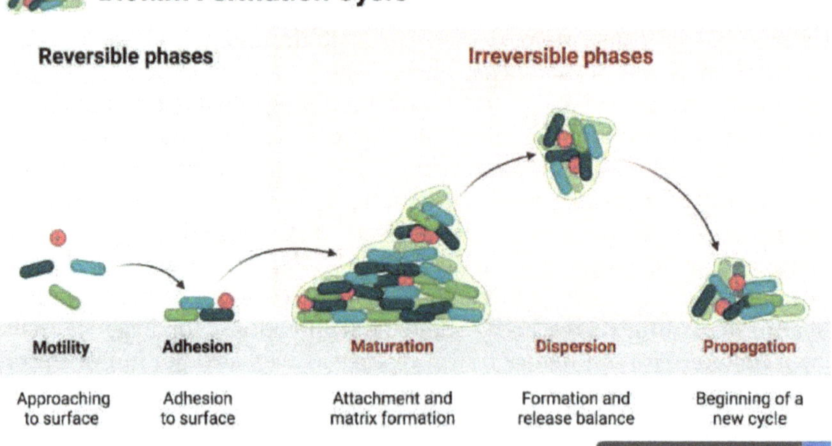

Figure 1. Biofilm formation cycle showing reversible and irreversible phases of the complex population in the microbial biofilm formation cycle.

Biofilms on industrial equipment reduce efficiency and cause corrosion. This can cause expensive industrial equipment failure and downtime. The worst aspects of biofilms are their capacity to cause illnesses and their potential to impede industrial efficiency. Their endurance and antibiotic resistance make them a menace to human health and industry. Biofilms in water treatment plants can remove pollutants but also breed hazardous infections.

Status of Biofilm Investigation in the Next Generation Sequencing (NGS) Era

Observing heaps of *Pseudomonas aeruginosa* cells in the sputum and the lung tissues of chronically infected cystic fibrosis patients in the early 1970s sparked the concept of biofilm and biofilm-associated infections, and their primacy in medicine (Hoiby et al. 1977). Shortly after the first two reports on biofilms (Jendresen and Glantz 1981, Jendresen et al. 1981), the term 'biofilm' was introduced to the medical society by J. W. Corterton in 1985 (Nickel et al. 1985). Over the course of 38 years, there have been a number of technical workshops and meetings deliberating on biofilms. In 2015, a set of guidelines for diagnosing and treating biofilm infections was released (Hoiby et al. 2015). With the development of high-throughput sequencing technologies, there have recently been numerous breakthroughs in the field of biofilm studies. These methods have made it possible to study biofilms in both fresh water and salt water. Understanding the active microbiome, gene expression patterns, and pathways in biofilm samples is possible thanks to the meta-transcriptomic analysis (Yao et al. 2021).

The hypervariable regions of the 16S rRNA gene are primarily sequenced using Illumina next-generation sequencing (NGS), which ensures the species identification in the biofilms (Papadatou et al. 2021). Using the Illumina HiSeq2500 platform, DNA libraries made from the isolated organisms' genomic DNA can be sequenced (Summers et al. 2022). The sequencing data is processed using QIIME with the default settings. By removing inferior reads, barcodes, primers and adaptors, it enables the generation of high-quality sequences. Operational taxonomic units (OTUs) are created from such high-quality sequences (Zhang et al. 2020). On the basis of OTUs, the GreenGene database is typically used to screen for species identification (Yao et al. 2021). It presents unique taxonomic profiles and patterns of bacterial species diversity in the biofilms (Antunes et al. 2020). However, compared to the GreenGene and RDP databases, the SILVA database with curated 16S rRNA gene sequences offers lower error rates (Papadatou et al. 2021).

The NGS techniques have been extensively used in studies that sought to identify the biofilms connected to various materials dumped in fresh and salt waters. Using Illumina NGS, which focused on the 16S rRNA gene, it has been possible to identify the differences in the marine biofilm microbiomes that were related to the fouling control systems (Papadatou et al. 2021). The high-throughput screening techniques can also be used to identify the biofilm community related to microbial fuel cells. As a result, a reliable method for identifying the prokaryotic and eukaryotic community in the fuel cell is established. To identify the strains, the OTUs were compared with the SILVA, GreenGene, and RDP databases (Zhang 2022). NGS of 16S rRNA gene amplicons can also be used to characterize the biofilms in harsh environments like the hot springs or permafrost. Gene amplification and sequencing using the MiSeqTM Personal Sequencing System technology have been accomplished using a two-stage PCR method combining the use of 515F and Pro-mod-805R primers with a dual-indexed Illumina TruSeq sequencing primer adapter (Kochetkova et al. 2020). Antibiotic resistance genes in biofilms are typically found using shotgun metagenomic methods in a variety of environmental samples. Using tools like HiSeq, samples are shotgun sequenced. Gene-host relationships in individual samples are examined by searching the contigs for phylogenetic clues and physical co-occurrences (Kneis et al. 2022). Metagenomic sequencing techniques can be used to analyze the biofilms on rock surfaces, and results have shown that biofilm age is the primary determinant of the composition of a biofilm (Summers et al. 2022). The aquatic rheophyte *Hanseniella heterophylla* microbiota has been studied using sequencing methods that combined paired-end Illumina sequencing of bacterial 16S rRNA and fungal nuclear ribosomal internal transcribed spacer region (Purahong et al. 2021). The NGS techniques offer a robust, effective and reliable way to analyze biofilm composition in a variety of environments, including contaminated water bodies, fresh and marine environments, and even man-made elements like fuel cells and fouling control systems.

Future Perspective – Human Microbiome after 100 Years (HM100)

As present, it is possible for people to successfully maintain a healthy gut microbiota in order to live a healthy life. However, the pace of research in fields such as establishing artificial microbiomes and microcosms is excruciatingly stagnant. There is some evidence that transplanting the microbiome of the gut can be an effective treatment. A number of researchers are focusing their efforts on microbiome transplantation or microbiome replacement processes. After a hundred years, it might be possible to restore a healthy gastrointestinal tract by repopulating the microbiome with healthy bacteria in order to treat all of the conditions that are brought on by a damaged microbiome. In addition, certain immunologically compatible and immune system-matched individuals may choose to have a transplant of their complete digestive tract, together with a healthy microbiome. This procedure is known as digestive track transplant. This will revolutionize human microbiome research and gut microbiome therapy.

Conclusion

The emerging field of next-generation sequencing has revolutionized the study of the gut microbiome, virome and biofilm investigations, providing unprecedented insights into the complex interactions between microorganisms and their hosts. The ability to analyze microbial diversity, community structure and functional capacity has revealed a wealth of information about the role of these microbial communities in human health and disease. Further research is needed to fully understand the complex interplay between host physiology, environmental factors and microbial communities to achieve the goal of HM100. Nonetheless, it is amply clear that the potential held by these approaches to transform our understanding of microbial communities and their impact on human health is significant.

Acknowledgement

BNR thanks Yenepoya (Deemed to be University) for funding this research (Project No.: YU/SeedGrant/104-2021).

References

Abeles, S. R. and Pride, D. T. (2014). Molecular bases and role of viruses in the human microbiome. J. Mol. Biol., 426(23): 3892–3906.

Al-Ishaq, R. K., Koklesova, L., Kubatka, P. and Büsselberg, D. (2022). Immunomodulation by gut microbiome on gastrointestinal cancers: Focusing on colorectal cancer. Cancers, 14(9): 2140. doi: 10.3390/cancers14092140.

Alaoui Mdarhri, H., Benmessaoud, R., Yacoubi, H., Seffar, L., Guennouni Assimi, H., Hamam, M., Boussettine, R., Filali-Ansari, N., Lahlou, F. A., Diawara, I., Ennaji, M. M. and Kettani-Halabi, M. (2022). Alternatives therapeutic approaches to conventional antibiotics: advantages, limitations and potential application in medicine. Antibiotics, 11(12): 1826.

Antunes, J. T., Sousa, A. G. G., Azevedo, J., Rego, A., Leão, P. N. and Vasconcelos, V. (2020). Distinct temporal succession of bacterial communities in early marine biofilms in a portuguese atlantic port. Frontiers in Microbiology, 11: 1938. https://doi.org/10.3389/fmicb.2020.01938.

Balabanova, L., Averianova, L., Marchenok, M., Son, O. and Tekutyeva, L. (2021). Microbial and genetic resources for cobalamin (Vitamin B12) biosynthesis: From ecosystems to industrial biotechnology. Int. J. Mol. Sci., 22(9): 4522.

Barone, M., Garelli, S., Rampelli, S., Agostini, A., Matysik, S., D'Amico, F., Krautbauer, S., Mazza, R., Salituro, N., Fanelli, F., Iozzo, P., Sanz, Y., Candela, M., Brigidi, P., Pagotto, U. and Turroni, S. (2022). Multi-omics gut microbiome signatures in obese women: Role of diet and uncontrolled eating behavior. BMC Med., 20(1): 500.

Begum, N., Mandhare, A., Tryphena, K. P., Srivastava, S., Shaikh, M. F., Singh, S. B. and Khatri, D. K. (2022). Epigenetics in depression and gut-brain axis: A molecular crosstalk. Front. Aging Neurosci., 14: 1048333.

Berg, G., Rybakova, D., Fischer, D., Cernava, T., Vergès, M. C., Charles, T., Chen, X., Cocolin, L., Eversole, K., Corral, G. H., Kazou, M., Kinkel, L., Lange, L., Lima, N., Loy, A., Macklin, J. A., Maguin, E., Mauchline, T., McClure, R., Mitter, B., Ryan, M., Sarand, I., Smidt, H., Schelkle, B., Roume, H., Kiran, G. S., Selvin, J., Souza, R. S. C., van Overbeek, L., Singh, B. K., Wagner, M., Walsh, A., Sessitsch, A. and Schloter, M. (2020). Microbiome definition re-visited: Old concepts and new challenges. Microbiome, 8(1): 103.

Blyton, M. D. J., Young, P. R., Moore, B. D. and Chappell, K. J. (2022). Geographic patterns of koala retrovirus genetic diversity, endogenization, and subtype distributions. PNAS, 119(33): e2122680119.

Borodovich, T., Shkoporov, A. N., Ross, R. P. and Hill, C. (2022). Phage-mediated horizontal gene transfer and its implications for the human gut microbiome. Gastroenterol. Rep., 10: goac012.

Breton, J., Galmiche, M. and Déchelotte, P. (2022). Dysbiotic gut bacteria in obesity: An overview of the metabolic mechanisms and therapeutic perspectives of next-generation probiotics. Microorganisms, 10(2): 452.

Campos-Madueno, E. I., Moradi, M., Eddoubaji, Y., Shahi, F., Moradi, S., Bernasconi, O. J., Moser, A. I. and Endimiani, A. (2023). Intestinal colonization with multidrug-resistant Enterobacterales: Screening, epidemiology, clinical impact, and strategies to decolonize carriers. Eur. J. Clin. Microbiol. Infect. Dis., 42: 229–254.

Cao, Z., Sugimura, N., Burgermeister, E., Ebert, M. P., Zuo, T. and Lan, P. (2022). The gut virome: A new microbiome component in health and disease. EBioMedicine, 81: 104113.

Carvalho, C. P., Ren, R., Han, J. and Qu, F. (2023). Natural selection, intracellular bottlenecks of virus populations, and viral superinfection exclusion. Annu. Rev. Virol., 9: 121–37.

Chandra, P., Grigsby, S. J. and Philips, J. A. (2022). Immune evasion and provocation by Mycobacterium tuberculosis. Nat. Rev. Microbiol., 20: 750–766.

Conlon, M. A. and Bird, A. R. (2014). The impact of diet and lifestyle on gut microbiota and human health. Nutrients, 7(1): 17–44.

Cronin, P., Joyce, S. A., O'Toole, P. W. and O'Connor, E. M. (2021). Dietary fibre modulates the gut microbiota. Nutrients, 13(5): 1655.

Dailey, H. A., Dailey, T. A., Gerdes, S., Jahn, D., Jahn, M., O'Brian, M. R. and Warren, M. J. (2017). Prokaryotic heme biosynthesis: Multiple pathways to a common essential product. Microbiol. Mol. Biol. Rev., 81(1): e00048-16.

Dethlefsen, L., McFall-Ngai, M. and Relman, D. A. (2007). An ecological and evolutionary perspective on human-microbe mutualism and disease. Nature, 449(7164): 811–818. https://doi.org/10.1038/nature06245.

Dong, J., Wang, W., Zhou, W., Zhang, S., Li, M., Li, N., Pan, G., Zhang, X., Bai, J. and Zhu, C. (2022). Immunomodulatory biomaterials for implant-associated infections: From conventional to advanced therapeutic strategies. Biomater. Res., 26: 72.

Erttmann, S. F., Swacha, P., Aung, K. M., Brindefalk, B., Jiang, H., Härtlova, A., Uhlin, B. E., Wai, S. N. and Gekara, N. O. (2022). The gut microbiota prime systemic antiviral immunity via the cGAS-STING-IFN-I axis. Immun., 55(5): 847–861.e10.

Fulci, V., Stronati, L., Cucchiara, S., Laudadio, I. and Carissimi, C. (2021). Emerging roles of gut virome in pediatric diseases. International Journal of Molecular Sciences, 22(8): 4127.

Gao, C., Xu, L., Montoya, L., Madera, M., Hollingsworth, J., Chen, L., Purdom, E., Singan, V., Vogel, J., Hutmacher, R. B., Dahlberg, J. A., Coleman-Derr, D., Lemaux, P. G. and Taylor, J. W. (2022b). Co-occurrence networks reveal more complexity than community composition in resistance and resilience of microbial communities. Nat. Commun., 13(1): 3867.

Gao, Y., Lu, Y., Dungait, J. A. J., Liu, J. B., Lin, S. H., Jia, J. J. and Yu, G. R. (2022a). The "regulator" function of viruses on ecosystem carbon cycling in the anthropocene. Front. Public Health, 10: 858615.

Gibiino, G., De Siena, M., Sbrancia, M., Binda, C., Sambri, V., Gasbarrini, A. and Fabbri, C. (2021). Dietary habits and gut microbiota in healthy adults: focusing on the right diet: A systematic review. Int. J. Mol. Sci., 22: 6728.

Gilbert, J. A., Blaser, M. J., Caporaso, J. G., Jansson, J. K., Lynch, S. V. and Knight, R. (2018). Current understanding of the human microbiome. Nat. Med., 24(4): 392–400.

Godlewska, U., Bulanda, E. and Wypych, T. P. (2022). Bile acids in immunity: Bidirectional mediators between the host and the microbiota. Front. Immunol., 13: 949033.

Goettsch, W., Beerenwinkel, N., Deng, L., Dölken, L., Dutilh, B. E., Erhard, F., Kaderali, L., von Kleist, M., Marquet, R., Matthijnssens, J., McCallin, S., McMahon, D., Rattei, T., Van Rij, R. P., Robertson, D. L., Schwemmle, M., Stern-Ginossar, N. and Marz, M. (2021). ITN—VIROINF: Understanding (harmful) virus-host interactions by linking virology and bioinformatics. Viruses, 13: 766.

Gomaa, E. Z. (2020). Human gut microbiota/microbiome in health and diseases: A review. Antonie van Leeuwenhoek, 113(12): 2019–2040.

Grice, E. A. and Segre, J. A. (2012). The human microbiome: our second genome. Annu. Rev. Genomics Hum. Genet., 13: 151–170.

Gupta, A., Gupta, R. and Singh, R. L. (2016). Microbes and environment. Principles and Applications of Environmental Biotechnology for a Sustainable Future, 43–84.

Handelsman, J. (2004). Metagenomics: Application of genomics to uncultured microorganisms. Microbiol. Mol. Biol. Rev., 68(4): 669–685.

He, Z., Qin, L., Xu, X. and Ding, S. (2022). Evolution and host adaptability of plant RNA viruses: Research insights on compositional biases. Comput. Stru. Biotechnol. J., 20: 2600–2610. doi: 10.1016/j.csbj.2022.05.021.

Hoiby, N., Bjarnsholt, T., Moser, C., Bassi, G. L., Coenye, T., Donelli, G., Hall-Stoodley, L., Holá, V., Imbert, C., Kirketerp-Møller, K., Lebeaux, D., Oliver, A., Ullmann, A. J., Williams, C. and ESCMID Study Group for Biofilms and Consulting External Expert Werner Zimmerli. (2015). ESCMID guideline for the diagnosis and treatment of biofilm infections 2014. Clin. Microbiol. Infect., 21(1): S1–S25.

Hoiby, N., Flensborg, E. W., Beck, B., Friis, B., Jacobsen, S. V. and Jacobsen, L. (1977). *Pseudomonas aeruginosa* infection in cystic fibrosis. Diagnostic and prognostic significance of *Pseudomonas aeruginosa* precipitins determined by means of crossed immunoelectrophoresis. Scand. J. Respir. Dis., 58(2): 65–79.

Hou, K., Wu, Z. X., Chen, X. Y., Wang, J. Q., Zhang, D., Xiao, C., Zhu, D., Koya, J. B., Wei, L., Li, J. and Chen, Z. S. (2022). Microbiota in health and diseases. Signal Transduct. Target Ther., 7(1): 135.

Iljazovic, A., Roy, U., Gálvez, E. J. C., Lesker, T. R., Zhao, B., Gronow, A., Amend, L., Will, S. E., Hofmann, J. D., Pils, M. C., Schmidt-Hohagen, K., Neumann-Schaal, M. and

Strowig, T. (2021). Perturbation of the gut microbiome by *Prevotella* spp. enhances host susceptibility to mucosal inflammation. Mucosal. Immunol., 14(1): 113–124.

Indira, M., Venkateswarulu, T. C., Abraham Peele, K., Nazneen Bobby, M. and Krupanidhi, S. (2019). Bioactive molecules of probiotic bacteria and their mechanism of action: A review. 3 Biotech., 9(8): 306.

Irwin, N. A. T., Pittis, A. A., Richards, T. A. and Keeling, P. J. (2022). Systematic evaluation of horizontal gene transfer between eukaryotes and viruses. Nat. Microbiol., 7: 327–336.

Jendresen, M. D. and Glantz, P. O. (1981). Clinical adhesiveness of selected dental material. An *in vivo* study. Acta Odontol. Scand., 39: 39–45.

Jendresen, M. D., Glantz, P. O., Baier, R. E. and Eick, J. D. (1981). Microtopography and clinical adhesiveness of an acid etched tooth surface. An *in vivo* study. Acta Odontol. Scand., 39: 47–53.

Kato, S., Nagasawa, T., Uehara, O., Shimizu, S., Sugiyama, N., Hasegawa-Nakamura, K., Noguchi, K., Hatae, M., Kakinoki, H. and Furuichi, Y. (2022). Increase in Bifidobacterium is a characteristic of the difference in the salivary microbiota of pregnant and non-pregnant women. BMC Oral Health, 22(1): 260.

Kneis, D., Berendonk, T. U., Forslund, S. K. and Hess, S. (2022). Antibiotic resistance genes in river biofilms: A metagenomic approach toward the identification of sources and candidate hosts. Environmental Science & Technology, 56(21): 14913–14922. https://doi.org/10.1021/acs.est.2c00370.

Kochetkova, T. V., Toshchakov, S. V., Zayulina, K. S., Elcheninov, A. G., Zavarzina, D. G., Lavrushin, V. Y., Bonch-Osmolovskaya, E. A. and Kublanov, I. V. (2020). Hot in cold: microbial life in the hottest springs in permafrost. Microorganisms, 8(9): 1308. https://doi.org/10.3390/microorganisms8091308.

Lal, A., Kil, E. J., Vo, T. T. B., Wira Sanjaya, I. G. N. P., Qureshi, M. A., Nattanong, B., Ali, M., Shuja, M. N. and Lee, S. (2022). Interspecies recombination-led speciation of a novel *Geminivirus* in pakistan. Viruses, 14(10): 2166.

LaTourrette, K. and Garcia-Ruiz, H. (2022). Determinants of virus variation, evolution, and host adaptation. Pathogens, 11(9): 1039.

Lean, C. H., Doolittle, W. F. and Bielawski, J. P. (2022). Community-level evolutionary processes: Linking community genetics with replicator-interactor theory. PNAS, 119(46): e2202538119.

Leeming, E. R., Johnson, A. J., Spector, T. D. and Le Roy, C. I. (2019). Effect of diet on the gut microbiota: rethinking intervention duration. Nutrients, 11(12): 2862.

Leeuwendaal, N. K., Stanton, C., O'Toole, P. W. and Beresford, T. P. (2022). Fermented foods, health and the gut microbiome. Nutrients, 14(7): 1527.

Limoli, D. H., Jones, C. J. and Wozniak, D. J. (2015). Bacterial extracellular polysaccharides in biofilm formation and function. Microbiol. Spectr., 3(3): 10.1128/microbiolspec.MB-0011-2014.

Lin, K., Zhu, L. and Yang, L. (2022). Gut and obesity/metabolic disease: Focus on microbiota metabolites. MedComm., 3(3): e171.

Martín, R., Miquel, S., Langella, P. and Bermúdez-Humarán, L. G. (2014). The role of metagenomics in understanding the human microbiome in health and disease. Virulence, 5(3): 413–423.

Matijašić, M., Meštrović, T., Paljetak, H. Č., Perić, M., Barešić, A. and Verbanac, D. (2020). Gut microbiota beyond bacteria-mycobiome, virome, archaeome, and eukaryotic parasites in IBD. Int. J. Mol. Sci., 21(8): 2668.

Miller, T. L., Wolin, M. J., Conway de Macario, E. and Macario, A. J. (1982). Isolation of Methanobrevibacter smithii from human feces. Applied and Environmental Microbiology, 43(1): 227–232. https://doi.org/10.1128/aem.43.1.227-232.1982.

Moughaizel, M., Dagher, E., Jablaoui, A., Thorin, C., Rhimi, M., Desfontis, J. C. and Mallem, Y. (2022). Long-term high-fructose high-fat diet feeding elicits insulin resistance, exacerbates dyslipidemia and induces gut microbiota dysbiosis in WHHL rabbits. PLoS ONE, 17(2): e0264215.

Mousa, W. K., Chehadeh, F. and Husband, S. (2022). Microbial dysbiosis in the gut drives systemic autoimmune diseases. Front. Immun., 13: 906258.

Mukhopadhya, I., Segal, J. P., Carding, S. R., Hart, A. L. and Hold, G. L. (2019). The gut virome: The 'missing link' between gut bacteria and host immunity? Therapeutic Adv. Gastroenterol., 12: 1756284819836620.

Murphy, E. A., Velazquez, K. T. and Herbert, K. M. (2015). Influence of high-fat diet on gut microbiota: a driving force for chronic disease risk. Curr. Opin. Clin. Nutr. Metab. Care., 18(5): 515–520.

National Research Council (US). (2004). Committee to update science, medicine, and animals. science, medicine, and animals. Washington (DC): National Academies Press (US). doi: 10.17226/10733.

Nickel, J. C., Ruseska, I., Wright, J. B. and Costerton, J. W. (1985). Tobramycin resistance of *Pseudomonas aeruginosa* cells growing as a biofilm on urinary catheter material. Antimicrob. Agents Chemother., 27(4): 619–624.

Nogal, A., Valdes, A. M. and Menni, C. (2021). The role of short-chain fatty acids in the interplay between gut microbiota and diet in cardio-metabolic health. Gut Microbes, 13(1): 1–24. https://doi.org/10.1080/19490976.2021.1897212.

Opal, S. M. (2009). A brief history of microbiology and immunology. Vaccines: A Biography, 31–56.

Papadatou, M., Robson, S. C., Dobretsov, S., Watts, J. E. M., Longyear, J. and Salta, M. (2021). Marine biofilms on different fouling control coating types reveal differences in microbial community composition and abundance. MicrobiologyOpen, 10: e1231. https://doi.org/10.1002/mbo3.1231.

Pisani, A., Rausch, P., Bang, C., Ellul, S., Tabone, T., Marantidis Cordina, C., Zahra, G., Franke, A. and Ellul, P. (2022). Dysbiosis in the gut microbiota in patients with inflammatory bowel disease during remission. Microbiol. Spectr., 10(3): e0061622.

Prasoodanan, P. K. V., Sharma, A. K., Mahajan, S., Dhakan, D. B., Maji, A., Scaria, J. and Sharma, V. K. (2021). Western and non-western gut microbiomes reveal new roles of Prevotella in carbohydrate metabolism and mouth-gut axis. NPJ Biofilms and Microbiomes, 7(1): 77.

Prosser, J. I., Bohannan, B. J., Curtis, T. P., Ellis, R. J., Firestone, M. K., Freckleton, R. P., Green, J. L., Green, L. E., Killham, K., Lennon, J. J., Osborn, A. M., Solan, M., van der Gast, C. J. and Young, J. P. (2007). The role of ecological theory in microbial ecology. Nature Rev. Microbiol., 5(5): 384–392.

Purahong, W., Mapook, A., Wu, Y. T. and Chen, C. T. (2019). Characterization of the castanopsis carlesii deadwood mycobiome by pacbio sequencing of the full-length fungal nuclear ribosomal internal transcribed spacer (ITS). Front Microbiol., 10: 983. doi: 10.3389/fmicb.2019.00983.

Radjabzadeh, D., Bosch, J. A., Uitterlinden, A. G., Zwinderman, A. H., Ikram, M. A., van Meurs, J. B. J., Luik, A. I., Nieuwdorp, M., Lok, A., van Duijn, C. M., Kraaij, R. and Amin, N. (2022). Gut microbiome-wide association study of depressive symptoms. Nat. Commun., 13(1): 7128.

Radman, M. (2022). Speciation of genes and genomes: Conservation of DNA polymorphism by barriers to recombination raised by mismatch repair system. Front. Gen., 13: 803690.

Repac Antić, D., Parčina, M., Gobin, I. and Petković Didović, M. (2022). Chelation in antibacterial drugs: From nitroxoline to cefiderocol and beyond. Antibiotics, 11(8): 1105. doi: 10.3390/antibiotics11081105.

Richert-Pöggeler, K. R., Iskra-Caruana, M. L. and Kishima, Y. (2022). Editorial: DNA virus and host plant interactions from antagonism to endogenization. Front. Plant Sci., 13: 1014516.

Rowland, I., Gibson, G., Heinken, A., Scott, K., Swann, J., Thiele, I. and Tuohy, K. (2018). Gut microbiota functions: Metabolism of nutrients and other food components. European J. Nutri., 57(1): 1–24.

Sadrekarimi, H., Gardanova, Z. R., Bakhshesh, M., Ebrahimzadeh, F, Yaseri, A. F., Thangavelu, L., Hasanpoor, Z., Zadeh, F. A. and Kahrizi, M. S. (2022). Emerging role of human microbiome in cancer development and response to therapy: Special focus on intestinal microflora. J. Transl. Med., 20: 301.

Siddiqui, R., Boghossian, A., Alharbi, A. M., Alfahemi, H. and Khan, N. A. (2022). The pivotal role of the gut microbiome in colorectal cancer. Biol., 11: 1642.

Slotten, H. R. (2014). The Oxford Encyclopedia of the history of American Science, Medicine, and Technology. Oxford University Press. ISBN: 9780199766666.

Somovilla, P., Rodríguez-Moreno, A., Arribas, M., Manrubia, S. and Lázaro, E. (2022). Standing genetic diversity and transmission bottleneck size drive adaptation in bacteriophage Qβ. Int. J. Mol. Sci., 23(16): 8876.

Spencer, L., Olawuni, B. and Singh, P. (2022). Gut virome: Role and distribution in health and gastrointestinal diseases. Front. Cell. Infect. Microbiol., 12: 836706.

Summers, K. M., Bush, S. J., Wu, C. and Hume, D. A. (2022). Generation and network analysis of an RNA-seq transcriptional atlas for the rat. NAR Genom. Bioinform., 4(1): lqac017. doi: 10.1093/nargab/lqac017.

Sutton, T. D. S. and Hill, C. (2019). Gut bacteriophage: Current understanding and challenges. Front. Endocrinol., 10: 784.

Thakur, A., Mikkelsen, H. and Jungersen, G. (2019). Intracellular pathogens: Host immunity and microbial persistence strategies. J. Immunol. Res., 2019: 1356540.

Tiamani, K., Luo, S., Schulz, S., Xue, J., Costa, R., Khan Mirzaei, M. and Deng, L. (2022). The role of virome in the gastrointestinal tract and beyond. FEMS Microbiology Reviews, 46(6): fuac027.

Tomova, A., Bukovsky, I., Rembert, E., Yonas, W., Alwarith, J., Barnard, N. D. and Kahleova, H. (2019). The Effects of vegetarian and vegan diets on gut microbiota. Front. Nutr., 6: 47.

Ungaro, F., Massimino, L., D'Alessio, S. and Danese, S. (2019). The gut virome in inflammatory bowel disease pathogenesis: From metagenomics to novel therapeutic approaches. United European Gastroenterology Journal, 7(8): 999–1007.

Vera-Urbina, F., Dos Santos-Torres, M. F., Godoy-Vitorino, F. and Torres-Hernández, B. A. (2022). The gut microbiome may help address mental health disparities in hispanics: A narrative review. Microorganisms, 10(4): 763.

Wang, L., Wang, S., Zhang, Q., He, C., Fu, C. and Wei, Q. (2022). The role of the gut microbiota in health and cardiovascular diseases. Mol. Biomed., 3(1): 30.

Watane, A., Raolji, S., Cavuoto, K. and Galor, A. (2022). Microbiome and immune-mediated dry eye: A review. BMJ Open Ophthalmol., 7: e000956.

Willis, J. R. and Gabaldón, T. (2020). The Human oral microbiome in health and disease: From sequences to ecosystems. Microorganisms, 8(2): 308.

Yao, L., Li, X., Zhou, Z., Shi, D., Li, Z., Li, S., Yao, H., Yang, J., Yu, H. and Xiao, Y. (2021). Age-Based Variations in the Gut Microbiome of the Shennongjia (Hubei) Golden Snub-Nosed Monkey (Rhinopithecus roxellana hubeiensis). BioMed Research International, 2021: 6667715. https://doi.org/10.1155/2021/6667715.

Yao, S., Hao, L., Zhou, R., Jin, Y., Huang, J. and Wu, C. (2022). Formation of biofilm by tetragenococcus halophilus benefited stress tolerance and anti-biofilm activity against *S. aureus* and *S. Typhimurium*. Frontiers in Microbiology, 13: 819302. https://doi.org/10.3389/fmicb.2022.819302.

Ye, J., Wu, Z., Zhao, Y., Zhang, S., Liu, W. and Su, Y. (2022). Role of gut microbiota in the pathogenesis and treatment of *Diabetes mullites*: Advanced research-based review. Front. Microbiol., 13: 1029890.

Zhang, L., Jiang, X., Li, A., Waqas, M., Gao, M., Li, K., Xie, G., Zhang, J., Mehmood, K., Zhao, S., Wangdui, B. and Li, J. (2020). Characterization of the microbial community structure in intestinal segments of yak (Bos grunniens). Anaerobe, 61: 102115. https://doi.org/10.1016/j.anaerobe.2019.102115.

Zhang, P. (2022). Influence of foods and nutrition on the gut microbiome and implications for intestinal health. Int. J. Mol. Sci., 23: 9588. doi: 10.3390/ijms23179588.

Zheng, D., Liwinski, T. and Elinav, E. (2020). Interaction between microbiota and immunity in health and disease. Cell Research, 30: 492–506.

Zhu, F., Tu, H. and Chen, T. (2022). The Microbiota-gut-brain axis in depression: The potential pathophysiological mechanisms and microbiota combined antidepression effect. Nutrients, 14(10): 2081.

Index

P

S

T

V

For Product Safety Concerns and Information please contact our EU representative GPSR@taylorandfrancis.com
Taylor & Francis Verlag GmbH, Kaufingerstraße 24, 80331 München, Germany

www.ingramcontent.com/pod-product-compliance
Lightning Source LLC
LaVergne TN
LVHW010558110826
845149LV00003B/697

* 9 7 8 1 0 3 2 5 0 6 0 6 7 *